The R&D Game:

Technical Men, Technical Managers, and Research Productivity

The R&D Game:

The M.I.T. Press

Technical Men, Technical Managers, and Research Productivity

David Allison, Editor

Massachusetts Institute of Technology
Cambridge, Massachusetts, and London, England

Set in Linotype Caledonia and printed and bound
in the United States of America by The Colonial Press Inc.

Designed by Dwight E. Agner.

Second Printing, September 1969

SBN 262 01025 9 (hardcover)

Library of Congress catalog card number: 68-27422

Contents

Part I The Man

Part I

The Man

Introduction

On the road to the New Jersey seashore, about sixty miles from New York City, you can sometimes catch a glimpse of a shiny box that sits over on the next hill. The foliage is thick and healthy in that part of the country, so you must keep a sharp eye or you will whiz by and never even notice the box, which is maybe a mile away. If you turn off the highway and go west, you see that the shiny object is actually a rather imposing building, five or six stories tall and about as wide as a large stadium. You cannot discern immediately what sort of structure you are looking at, whether it is some sort of colossal insurance office or perhaps even a benign penal institution, but if you explore the matter just a little further you discover it to be the workplace of several thousand scientists and engineers.

There is nothing remarkable about finding large clusters of scientists and engineers working amid bucolic surroundings. If one knows where to look, he can ferret out many such colonies around San Diego and Los Angeles, along the San Francisco Bay and the shores of Lake Michigan, outside Boston, Dallas, and many other cities. Perhaps the most remarkable thing is their ubiquity.

But when I look at one of these huge workplaces of science, a series of thoughts comes to mind. First is its size and potential: If science and technology are truly the instruments of change we believe them to be, if ours is indeed the Age of Science or the Age of Technology, then this structure has significance beyond its size or population; perhaps it is, as

Pasteur would have described it, a temple of the future. But then the thought occurs: Is this really so? Is it not true that for every scientist in this particular laboratory, there are a thousand others elsewhere, from London to Tokyo, from Leningrad to Johannesburg, from Seattle to Buenos Aires? And here one must concede that this single building, however large or impressive it may appear, is one drop in a very large bucket. After all, there are four or five million scientists and engineers in the world today, and no single laboratory, however distinguished, is more than an octave of the vast spectrum of world science. But there is one concluding thought in my series that always brings a sense of unquenchable wonder: When Shakespeare was working at the Globe, when Galileo was discovering the satellites of Jupiter, when John Smith was sailing to Virginia, there were fewer scientists in all the world than there are today in this single building.

But what do these numbers mean? How do we account for the fact that there are so many technical people in the world today? Some say it is simply a case of demand working its spell over supply. Jack Morton describes it as "technological ecology," as man's adaptive response to his changing environment. But how do we account for the *rate* at which this change is occurring? For example, someone has calculated that if present trends continue there will be more scientists in the world by the year 2040 than there will be people! What sort of environment is man adapting *to?*

I do not know the answer to that question, but one thing that is suggested by these unfathomable numbers is that perhaps we ought to try to see what these people are all about. Why do they want to be scientists anyway? — or engineers? What are they like? Why do they suddenly seem to be so numerous? Why, despite the accelerating rate at which the world is producing them, are there usually too few scientists and engineers to go around? Why are they so expensive?

The matter of expense troubles many people, including many of the scientists themselves. Particularly, as one might expect, the high cost of supporting science and technology troubles those who must get up the money, from the president of the little company, who has decided he must have a research laboratory, to the U.S. Congressman, who tries to be an intelligent guardian for the most generous benefactor science has known in the history of the world, the United States government. To each of these men, the company president and the U.S. Congressman, science seems to have an insatiable appetite. This is not simply an illusion of the penny-pinching bookkeeper. If you are that company president, you can almost know that the cost of running your research facility will double during the next decade. If you are the Congressman, you have watched federal expenditures for science and technology rise even faster; in recent times the cost has tripled every ten years.

This imposes a special responsibility upon such men, and indeed upon all men who are involved in the support and management of science and technology. If they are to be intelligent guaridans of the public purse or the company treasury — if they are to be effective supporters of science and technology — it is essential that they *understand innovation*. They must understand, for example, that innovation is not simply a flash of genius, not just a brilliant idea, but that it is a chain of events that stretches from the idea to something tangible and socially valuable. One need not be a scientist or engineer to know how the innovation process works, but if that process truly *is* to work, in a company or in an entire nation, it is imperative that the supporters and managers of the process be something more than generous men of good will.

Albert Camus gave us the lesson with this passage in *The Plague*:

"The evil that is in the world always comes of ignorance, and good intentions may do as much harm as malevolence, if they lack understanding. On the whole, men are more good than bad; that, however, isn't the real point. But they are more or less ignorant, and it is this that we call vice or virtue."

In terms of this book, *virtue* means understanding the innovation process. Our objective is to describe that process and those technical men who help to sustain it. Why is this a worthwhile exercise, either for us who labor here to picture this phenomenon or for you who are about to consider our comments? I believe it to be so because such understanding can enable man to expand his knowledge and to apply that knowledge to human betterment. With such understanding, we learn, for example, that a nation's betterment does not depend upon the number of scientists it educates, nor upon the number of Nobel prizes its scientists may win. Neither does a corporation's growth depend solely upon comparable measures. What *does* derive from an understanding of the innovation process is knowledge of how science and engineering *can* contribute to the growth of a nation or the growth of a corporation. Indeed, science *can* be expensive and disruptive, as its supporters frequently complain, but it can also be rewarding, both to those who practice it and to those who support the enterprise, and when this occurs — when a disease is cured forever or a "miracle" fabric is developed — it seems not to have been expensive at all. It was science, then technology, that provided the world with radar, atomic power, and the Xerox copy, jet propulsion, television, and cellophane, man-made diamonds, Pyrex glass, and the mighty transistor, Teflon, nylon, and, of course, the great computers. Each of these familiar items, and many, many others we might name, is most certainly a product of modern science and technology in that none existed and was, perhaps, not even envisioned as recently as 40 or 50 years ago.

But these success stories of modern science and technology, these cliché names we trot out each time we want to feel good about our so-called scientific know-how, are really rather freakish examples. For every billion-dollar idea like nylon, which cost less than a million dollars to conceive, develop, and get ready for pilot production, there is another idea, buried somewhere with a long-since-forgotten name, that cost a lot more and never earned a nickel. And for every great laboratory in the land, whether greatness be measured by the quality of the technical ideas generated or by the dollars those ideas ultimately turned out to be worth, there are many more that are not great at all, not even good.

Why should this be so? This question troubles many United

States corporate executives, as well as many thoughtful Europeans who are concerned over the widening "technology gap" between themselves and the United States. An answer that is sometimes put forth attributes the cause to the technical men themselves, one corporate answer being that "our scientists" are not as good, or perhaps not as diligent, as "their scientists." But this does not really define the problem for many corporations, and one certainly cannot account for the rapid growth of the United States economy versus the economies of certain countries of Western Europe by claiming that the United States has better scientists, or even more scientists. A more reasonable explanation is that the problem is managerial rather than technical, which is to say it is not so much a question of whose technical people are better, but rather how effectively those technical talents are being used. And this brings us back to the earlier discussion of the innovation process and why that process must be understood by those who manage and support science and technology. I must add here that such understanding cannot be guaranteed to come forth from any specific page, nor dare we promise that you will experience a flash of insight when you have put the book away and gone to the kitchen for cocoa. But I do believe you will know more about scientists themselves: how they think and why they get stuck and what motivates them. And I believe you will have a clearer picture of that amorphous thing called the research environment: how it affects the creative process and why an idea can take root in one place but not in another. And, finally, I believe you will see how science and technology fit into those larger structures, the corporation and the nation, and how both science and technology can contribute to their growth.

As you walk through this technical territory, you will meet a rather heterogeneous assortment of folk, including one psychiatrist, four psychologists, a couple of engineers, a few scientists, several technical managers, at least one philosopher, and a couple of cartoonists who have come along lest we take ourselves too seriously. These are the men whose ideas and observations you will be encountering. I have asked them to join the party because, quite frankly, I do not know enough about the subject to write the book myself. But I should state here that these men have not been assembled so that you will hear "all about" technical management or so that you will know all sides of the argument that revolves about the question of how to manage technical people. In that sense, this is a one-sided book, for the authors express a rather consistent philosophy toward the question. This is not to say that each man agrees completely with every comment of each of the others, but I believe there is sufficient consistency from chapter to chapter to save you utter confusion. This suggests an underlying prejudice, and I must tell you that such a prejudice did indeed exert itself when we set out. In fact, it was I who exerted it when the authors were selected, for I picked them, then worked as their editor as we put to paper the things they wanted to say.

There is one member of the party whom you will never meet here. We regret his anonymity, for it was he who encouraged us to set forth with these essays: This is Robert Colborn, the chief editor of *International Science & Technology*. Each of the pieces in this book appeared originally and in somewhat different form in his magazine.

During the first third of the book, we are going to look at the man himself. When we say the man, we of course mean the technical man — the scientist or the engineer — but we also mean the scientist or engineer who happens to work in an industrial organization. As you read about the man, note how often the word *strain* appears in the accounts of his daily activities, the point being that his life is not an easy one. But do not be misled to think the life of the industrial scientist is necessarily possessed of more strain or conflict than that of his colleague who lives on the campus. The life of any technical man is one of strains and conflicts if it is any kind of life at all, for otherwise it is apt to be a life of ennui and emptiness. The point is not that strain should be abolished, but that it should come from the pressures inherent in the creative process, rather than from extraneous sources that serve only to thwart the process.

This does not mean that we are going to talk about good guys and bad guys in this book — good guys whose aim is to protect creative people and bad guys whose aim is to corrupt them. If you are searching for experts who will strengthen your case for "leaving creative people alone, so they can really *be* creative," I fear you will find little ammunition here. Complete freedom to pursue whatever one wants to pursue turns out *not* to be the most desirable condition for an industrial laboratory, either from the viewpoint of the corporate managers or from the viewpoint of the scientists themselves — and certainly it is not a desirable condition from the viewpoint of the society to be served by that laboratory. It is this lack of complete freedom that gives the industrial laboratory its unpopular (and frequently misunderstood) reputation among many academic scientists. But it is also this quality that makes the industrial laboratory productive of new knowledge and useful technology. Indeed it is this quality of less-than-total freedom — or freedom within a mission — that makes the industrial environment such an interesting subject for study. It is interesting, first of all, because it is a study of people, quite often extraordinary people, and how they behave and perform in what is most certainly a planned environment. Psychiatrist Lawrence Kubie tells later on that, in science and engineering, creativity demands a flexibility of the human personality that may be greater than that of almost any other occupation to which man can dedicate himself. This demand is particularly severe in the world of industrial science and engineering, and sometimes it produces schizophrenic problems when the professional objectives of the technical man conflict with the objectives of his organization. When this happens, you begin to learn some things about the flexibility of the technical people who work around the place. You also learn about the adaptability of the organization and its managers, and how skillfully the organization is being managed for change.

Let us look closer at this conflict. Kubie tells us: "The scientist and the engineer must first of all maintain mastery of an enormous body of rapidly growing data, yet, at the same time, must be as freely imaginative as the poet, the artist, or the musician." He must also possess the capacity to direct his imaginative flights to real goals, to test the degree to which they are consonant with the real world, and finally, to project them into the future for new uses. Few other occupations demand so much, psychiatrist Kubie tells us, and those who know of the technical world are likely to affirm this observation. But in the environment whose orientation is industrial, the conditions are perhaps even more stringent than those described for the technical world as a whole, for here there is that extra element called relevance, which must be included when describing reality. Not only must the technical man's imaginative flights carry him to ideas that are original yet solidly based; if he is really to get off the ground with a new industrial idea, the idea must also carry the promise of profitability.

For some technical people, this constraint is too severe. In the opinion of such people, it is the mission orientation of the industrial organization that is at the heart of the trouble; rather than knuckle under to such a single purpose, they abandon the industrial world or refuse to venture there in the first place. It is probably fair to say that about half the young scientists in the country have such fears of the industrial laboratory, for of all the doctoral candidates in the physical sciences in recent years, only about half have gone to work in industry. And even among these brave recruits, we can be certain there are those who go off tentatively, intending one day to return to academia. This is a major problem for our modern industrial world, for if many from each year's crop of newly trained scientists are reluctant to consider industrial careers, preferring instead to hang around the university for a year of postdoctoral study or to teach and do some academic research, then the industrial laboratory can scarcely expect to be the center of excellence it hopes to be or thinks it is.

There are cases in which the problem is assaulted with money: "If we need the best brains in the country, then, by God, we will go and buy 'em." But such good old American horse sense, or blunderbuss, seldom returns the investment, because it never really solves the problem. To state the problem in its simplest terms: If an industrial organization wants to attract good scientists, it must be able to demonstrate that it intends that these people *do* good science. And the most effective way to demonstrate this — also the cheapest way, in the long run — is to be in the ball game. But even here we are not ready to celebrate a happy ending, for now we are back where we were a moment ago, facing the perennial conflict between what the scientist wants to pursue and what his organization wants him to pursue. For this critical stage of the struggle, we shall pick up a new character, whom we shall call our hero, or the goat, depending upon how the story ends, for this individual will play an important part in that determination, since he is the technical manager. We will meet him time and again before this saga ends, but for now we see him as the lonely soul who must coax or lead or somehow get a batch of smart individualists to work on problems that are relevant to a specific industry. He turns up in Ray Hyman's essay on creativity, worriedly asking psychologist Hyman whether there is some incentive system or bonus plan that will motivate his technical people to be "more creative." This manager has a serious problem, for unless his people come forth with some profitable and patentable ideas, his company is likely to go right down the drain — and during the past year not one of his technical people has suggested a darned thing. The first question the psychologist asks him is "What do your technical people think of working under some kind of bonus system?" And the manager says, "I am embarrassed to say that I have never discussed the problem with them." Well, to make a long point sharper, on that same day the manager called his men together and told them that the company had to have new ideas or it would not survive, and within the month the ideas started to flow in.

I like the story because it illustrates one of the most valuable contributions to be made to technical management by people in the social sciences, which is to remind the managers that their biggest problems, now and evermore, are *people* problems and that the best way to solve such problems is by approaching the people as people, rather than as things to be manipulated by bastardized gimmicks from some social scientist's bag of tricks.

Social scientists are equipped to make many contributions to the world of science and technology, as you will discover when you have read the first third of this book. For example, if you consider yourself something of a dud at solving what turn out to be "easy" problems, read the essay by Ray Hyman and his fellow psychologist Barry Anderson at the end of this first section. Here you will find some guidance from experimental psychologists that may have applicability in your world. Another kind of guidance, this time from social psychologists, is also squeezed into this first series of essays. Here I refer to the work of Donald Pelz and Frank Andrews, who have made several major studies of a number of research organizations to try to discover what makes for a stimulating research atmosphere: What kinds of relations among scientists, what kinds of personal motivations, raise or lower scientific achievement? In their two essays, Pelz and Andrews look at these questions in terms of research *freedom* — is there a correlation between freedom and achievement? — and in terms of research *diversity*— should the man stick to one area and try to master it, or is it better if he involves himself in several kinds of things? So far as I know, nobody has explored these questions with the thoroughness of these two authors, and if you are the conscientious manager I think you are, you are going to stay with Pelz and Andrews until you get their various messages. I make this special plea because their efforts may be a bit harder to follow than some of the others, not out of any clumsiness with the language, certainly, but because there may seem to be an inordinate amount of statistical information involved in the essays, and this is always a little more jarring than pure English.

If you do not happen to be a scientist or an engineer or even one who tries to manage those often-gifted, often-unmanageable people, I owe you a special greeting and a word of gratitude for having come so far into this book. Irrelevant reading does have its rewards, however, and before you have progressed beyond these first few essays, I suspect you may discover that the comments of the authors have wider applicability than you had supposed at the outset. While the authors talk in terms of scientists and engineers, I believe many of their observations apply as well to artists, writers, indeed to most creative people. I have been struck by this as I have read these essays again after having been away from them for some while. One could say that this book only happens to be about technical people, but the ideas and observations expressed here apply to creative people in many fields. I believe this is no accident, for the creative scientist and the creative writer, for example, are likely to have more in common — more to talk about, more to gripe about, more to laugh about, more to cry about — than, say, the creative scientist and his plodding, unfortunate colleague, or the creative writer and the hack. That mysterious and rare commodity called creativity is the common bond, the gift each took with him when he started down the road. Only later, when the road began to divide, did one wander off toward music, another toward literature, and others toward the fields of engineering and science. Perhaps this oversimplifies the situation, for it suggests that technological innovation is no different from composing a symphony or writing a book. Of course it is different, if only because it involves many different kinds of creative specialists, rather than a single artist or writer, and this elevates the problem to a higher order of complexity. But I do believe these essays have relevance outside the world of science and technology, especially in organizations of people whose collective effort is supposed to yield a creative result.

David Allison

1 The Industrial Scientist

He sits in his hard, gray chair and stares at the steel partition. His room is B105, which he shares with another young physicist. Both joined the laboratory during the previous summer. As he swivels, aimlessly, alone in his two-man cubicle he tries to assess these past few months: What has he accomplished? How is he doing? Suddenly, in the awful stillness, he is overcome with depression. He is no good. He will never make it here. "Dear God, whatever brought me to this place?"

The question actually comforts him. It helps him see that the place is wrong and he is right. He should never have come. That was the big mistake. What a fool! But the mistake does not reflect upon him as a scientist. This is the saving grace. It only reflects upon him as a man — a brash, young man who would not listen to advice. If only he had listened to Weis and the others, his faculty men in graduate school. . . . They knew what it would be like.

"I'm just not cut out for industry. Nothing wrong with that." He can live with himself now. He recalls something Weis said, that night they all got drunk: "There are scientists and there are industrial scientists." And isn't that the point, really? He is not an *industrial* scientist. He is a *scientist,* goddammit, and proud of it!

Then other questions sweep through his mind, and he is troubled again: "If I am really too good for this place, why am I afraid of it? If I'm so good, why do I think I will never make it here?" When he tries to confront these questions, he begins to suspect that he has been fooling himself, that Weis and the others were naïve. He will agree that the men who work here, in the industrial laboratory, are different in some ways from the scientists he knew as a graduate student, but he has not discovered the difference that Weis had implied in his slur. These people are good, very good. He must concede that his periodic depressions grow partly from the fear that the best of them are better scientists than he will ever be. But there is another source of his depression — something that gnaws at him and bewilders him and makes him lose respect for these people. Indeed, the better they appear, as scientists, the greater is his scorn. How can he describe this? He thinks of two men: One is McCullough, here at the lab; the other is Walker, who had been his thesis adviser.

Each man, in his way, is a brilliant scientist. But each sees his role in science in totally opposite terms. For example, McCullough seems almost to have an engineer's approach to things: It is difficult to capture his interest in a problem or a discovery, however exciting it might appear to you, unless he can sense that the thing might have commercial value. His favorite word is "relevance." Professor Walker, on the other hand, could be caused to explode with joy when presented with almost any example of imaginative thought. It need not be physics; it might be poetry, or music, or even a move on a chess board. If he liked it, he would say, "Beautiful, *beau*tiful." It is this quest for beauty that drives Walker in his work. In his eyes, beauty and utility have no relationship with one another. In fact, he would often startle admiring visitors by saying: "If you can find any commerical utility in what I am doing, I will stop doing it at once."

As the months go by, our young physicist will probably make an accommodation — one way or another. Either he will flee the "profit-oriented" world of the industrial laboratory, and perhaps forever carry a bitterness toward that world, or he will stick out that first, agonizing year — perhaps to see a paper published or a patent issued. If he stays, it will be due in some measure to a modest success. But so long as he stays, if even for a lifetime, he will feel one or another kind of strain. This will forever be so because the world of industrial science is a tough and complex world. Indeed, one might describe it as a pressure-cooker world — where discovery for discovery's sake is not enough, where profit is one important measure of success. Like science itself, it is an unnatural world, a world of man's creation, and there are countless reasons why it should not work.

In point of fact, its failures outnumber its successes — perhaps by 30 or 40 to 1. Again and again, it is a repetition of Victor Borge's comic tale of the despairing inventor of soda pop, who invented 1-UP and 2-UP and so on, and who gave up the battle after his sixth failure. But when it works, it's cellophane. And that is the fuel that makes it go, causing American industry to invest about $7 billion per year in research and development. (The federal government spends about twice this amount for research and development, and a substantial share of this federal money flows into the private corporations for the support of work performed by industrial scientists and engineers.)

Because it is a tough, unnatural world — and a gambler's world — it fascinates a good many people, from the sociologists who study the research environment to the economists who study its performance, from the psychologists who want to help the scientist be more creative to the psychiatrists who wonder why a man's creativity so often stops, leaving the pitiful victim dead in his tracks. Many of these social scientists are university people, and they receive support for their research from the federal government. But others with similar skills are employees of private industry; these are the industrial explorers, whose function is to try to understand the motivations and frustrations of the scientists in their own organizations. Most of these industrial people are invisible outside their own organizations: They seldom publish the results of their research and quite often they are not even known to their counterparts in other companies. The reluctance toward publication is understandable, for a perceptive social scientist who lives within an industrial laboratory eventually becomes an expert on the weaknesses of his environment. This has been the nature of his professional training — to discover those weaknesses — and this is the knowledge that gives him high value to his company.

Another kind of person is easily as fascinated as the social scientists in what makes scientists tick — or not tick. This is the gambler, the top manager, the boss. He comes in a variety of forms. Nice guys and tough guys. Smart guys and dull guys. But he always comes with cash, for the house minimum in this game is $50,000, which is roughly the cost per year of having one scientist on the premises.

The gamblers are the people who really create the excitement — and the strain. When a company brings an idea out of its laboratory, takes it through development, and finally establishes it in a profitable market, you can always find a gambler in the act. A smart, tough, well-heeled gambler. When nothing is coming out of the laboratory, you can know that the scientists have gone, or the gamblers have gone, or one group or the other has grown old and cautious. All bets are off.

Any such situation is a source of strain to the scientists. On the one hand, there is the strain that comes from dreams — dreams of new ideas brought to life — and on the other is the strain of frustration and guilt. Let us look more closely at each of these.

There exists among some top managers a mystical faith in the power of the research mind. The managers who are buoyed by this faith are sometimes armed, as well, with bits of research lore — such as the lore of Wallace Carouthers and how he came down from Harvard with his store of knowledge of condensation polymers, and how, $787,000 later, Du Pont was ready to build its first pilot plant for nylon. And the lore of cellophane and how that wondrous viscose film had been made ready for production after only $32,000 had been expended.

These are the gambler managers — good and bad. But the good ones know there is more to the tales of great achievement than the mere casting of bread upon waters. They know, for example, that transistors or xerography or Teflon fry pans do not appear full-blown from the creative minds of men. To be sure, without the creativity of Shockley, Brattain, and Bardeen at Bell Labs, there might be no transistor; without Chester Carlson's imagination and zealotry there might be no xerography; without Roy Plunkett, who could see the significance of his research accident, there might well be no Teflon. But the good managers know, too, that a number of other people, with various skills, also contributed significantly to each of these important developments. The transistor is a case in point: Here was a deliberate and aggressive attempt to exploit the field of semiconductors. To accomplish this, it was necessary at the laboratory to develop competence in a number of technical areas. One was crystallography — because crystal structure is basic to this field. Another was the growth of large, nearly perfect single crystals, for without these the research people would be working in the dark. Materials purity was still another. All of these investigations went on within a research group whose annual budget was about $500,000 — a substantial fraction of which was aimed at semiconductor research. And yet, when the program was launched — and for two years thereafter — no one gave, or indeed *could* give, any assurance that profits would ever come from the investment. And after the transistor discovery itself was made, it was necessary to put the device to work — and again, this required aggressive work on the part of still more specialists.

So the good gambler knows it to be a high-risk game, with failure a possibility at every stage. But he knows too that he can minimize the risks, and one way to do this, insofar as it pertains to the laboratory scientists, is to optimize the strain that accompanies industrial research. Sociologist Simon Marcson, a student of the industrial environment, concedes that strain is inherent in the industrial laboratory and that perpetual harmony is unrealistic, but he believes that strain can be managed. Marcson declares that the management of strain in the industrial laboratory depends upon "the evolvement of a new authority system, incorporating a mixture of executive authority and colleague authority." (When you read Rensis

Likert's essay on supervision, in the next section, you will see how one such authority system might be established.)

Economist Richard Nelson, in a study sponsored by the RAND Corporation, draws a similar conclusion. He asks: "Who is to decide which scientists will work on which projects?" When those projects are selected because they show "scientific promise," says Nelson, then the decision-making structure must be *decentralized.* He focused his study on one of the classic stories of modern industrial life — the development of the transistor. Because it illustrates *how* a group of shrewd research gamblers behaved, and postulates reasons for their success, let us hear some of Nelson's observations:

- The inventive activity that led to the transistor was similar in kind to the research work that led to nylon, Dacron, hybrid corn, and radar. This effort was *not* motivated by or directed toward the objective of a closely defined marketable product. The scientists had an idea of the practical fruits to which their work might lead, but the research project undertaken was far less closely focused on a single objective than were the projects that led to the jet engine, Kodachrome, or the automatic cotton picker — where the scientists and engineers were aiming at one particular target.
- A number of people, with different skills, contributed to the research, but the exact nature of the interactions among them could not have been predicted or planned in advance. Bardeen's analysis of surface states and Brattain's experimental skill were key factors, but nobody, at the start of the program, could have predicted that these particular experiments would be performed. Throughout the program, there was a great deal of informal exchange of ideas, and quite a bit of cooperation on experimental work. The major requirements for effective interaction seem to have been easy communication and the ability of individuals to drop what they were doing to help with problems their colleagues brought up, if these problems seemed interesting and important. There is little evidence of closely programmed and directed teamwork, or of any requirement for it.

• The ingredients that seem to have played an important role in the success of the project include the close interaction of a group of topflight scientists, a great deal of freedom in the course of the research, and an extremely strong interest on the part of at least one member — Shockley — in inventing a practical device.

Let us add one further observation: The people at Bell Labs had a strong hunch that semiconductor physics was the most *relevant* area of science to explore. Also, they had a strong faith that *understanding* would eventually reveal how to get amplification. And finally, as they looked at the theory and experiment, they had a sharp eye for *opportunity*. (You will hear more about the development of the transistor in Jack Morton's essay in the third section.)

We should ask ourselves a question at this point. The question, though it may seem frivolous, is nonetheless an important question to ask, in order to understand why there are productive scientists in some industrial laboratories and nonproductive, frustrated scientists in others. The question is this: How did these scientists *happen* to invent the device they did — and how did they happen *not to invent* something, however interesting it might be, that would have no value to their company? (Such things do happen, for the brilliant but unwanted discovery is one of the perennial hazards of industrial research.) The answer to the question really goes to the favorite word of our fictional friend, McCullough, earlier on: the word "relevance." James Fisk, the president of Bell Telephone Laboratories, expands this as follows: "Among a thousand scientific problems, a hundred or so will be interesting, but only one or two will be truly rewarding — both to the world of science and to us. What we try to provide is the atmosphere that will make selecting the one or two in a thousand a matter of individual responsibility and essentially automatic." And Nelson, in his study of the transistor, expands the answer still further by pointing out that there are no geneticists at Bell Labs, nor are there many organic chemists. (This was true when Nelson made the comment, but not today: Late in 1966, Bell Labs set up a new department of biophysics. One of its functions is to study information storage in RNA and DNA molecules.) "While the choice of the research area of the scientists in the research department is seldom subject to strong executive pressure," he says, "formal executive decisions strongly affect the allocation of research effort through the hiring process."

But the act of hiring, of itself, does not assure that the work of *all* the scientists in an industrial laboratory will be relevant to the goals of the company. The fact that a young solid-state physicist is hired for his proficiency and promise in that field is no guarantee that he will continue to be absorbed by the solid state. What does one do, four or five years later, if this young man says he would now like to turn his talents toward something new?

In most industrial laboratories, the answer is firm and businesslike. At Bell Labs, for example, vice president Jack Morton will say: "If a scientist in a communications laboratory comes along and says he wants to do work on drugs, you may have to suggest that there are other places where he would fit in better. You don't do basic industrial research because you think it is fashionable." At Du Pont, vice president Robert Hershey will say: "Research *per se* is not a suitable objective for an industrial organization. Research and its application, taken together and viewed as inseparable, are the legitimate goal." At IBM, vice president E. R. Piore will say: "There is not much free money available for research gambles, not as much as one would like to see. The whole process of research allocation is designed to protect our present inventory and our present line of products and to make sure that our future products will dominate the market. In this sense, allocation is dominated by profit considerations."

There may be exceptions. There may be industrial laboratories in which the scientist is free to pursue whatever he wishes for soever long as he wishes. Indeed, there are young scientists in industry who feel they have this freedom — and who may continue to feel it for a lifetime, so long as their interests happen to coincide with the interests of their organizations. But it is simply not true that the industrial environment is as permissive as the environment of a university laboratory — despite some industrial claims to the contrary. To try to preserve this belief, it seems to me, is to ignore the greatest capability that the industrial laboratory possesses — the ability to exploit new knowledge.

What is required for such exploitation — and why is the industrial environment uniquely equipped for the task? Let us look at four requirements. The first is talent — talent in sufficient numbers and of sufficient disciplinary diversity to attack a broad spectrum of problems. The large industrial laboratory can meet this requirement, but so too can the faculties of most good universities. A second requirement is an intimacy with the world of science. And again, both the university and the good industrial laboratory share this quality, for both will be involved in basic research. A third requirement is a scientific sophistication — a sense of what is intellectually promising and what should be explored. And here too, each has this ability. But the fourth requirement belongs mostly to industry: This is the ability to recognize and generate relevant advances in science *and* to be able to bring together the various talents of scientists and engineers who will carry the advance through research, development, manufacture, and finally to market. To be sure, such university laboratories as the Lincoln Laboratory of MIT or the Jet Propulsion Laboratory of Cal Tech have this capability. But these are not university facilities in an academic sense; rather, they are "mission-oriented" organizations, administered by the schools for one or another agency of the federal government. A university faculty is quite another matter. To expect a faculty to work cooperatively toward the practical development of a process or device (except in a period of national emergency) is to forget the words of Robert Hutchins, who once described the modern university as a series of separate schools and departments held together by a central heating system.

Perhaps it is little wonder that the world of the industrial scientist is laden with strains and perplexities. Science itself, in any setting, is an unsettling activity, and industrial science compounds this disorder by limiting its quest to *relevant* discoveries — discoveries that show promise of commercial payoff. Then it pushes hard to achieve that payoff. Moreover, industrial science is rather a new phenomenon. Where modern science enjoys a rich and heroic 400-year history, going back to Kelvin and Newton and ultimately to Galileo, industrial science spans only 90 years, to Edison's laboratory at Menlo Park.

It has created its own heroes, including Irving Langmuir of the General Electric Research Laboratory, whose work in surface chemistry won him the Nobel prize for chemistry in 1932 — the first American industrial scientist to be so honored — and Clinton Davisson of Bell Labs, who shared the prize for physics in 1937 with Britain's G. P. Thompson, and the trio at Bell Labs a generation later — Shockley, Brattain, and Bardeen — who won the prize for physics in 1956. But still, among young graduates of today, there is a reluctance toward entering the industrial world — perhaps the same reluctance Willis Whitney felt in 1900, when it was proposed that he leave the MIT faculty to establish a research laboratory at General Electric. Whitney accepted and went on to head the laboratory for thirty years, but he was most unwilling to abandon the academic world altogether, and for the first three years he kept his anchor near MIT and commuted between Cambridge and Schenectady.

Perhaps today's young scientists are a bit more willing to pull up stakes and see what the industrial world is like: Nearly half of each year's crop of physical science PhD's go to work in industry these days. But many do so with doubts and many, like our young physicist at the outset of this essay, suffer with those doubts for a long time thereafter. Why is this so? Let us seek an answer from a social scientist who has lived in the industrial research environment for a dozen years and is recognized as one of its keenest observers. This is Lowell Steele of G.E.:

For the young individual first coming into an industrial laboratory, the major source of stress is simply uncertainty. He comes with a whole list of questions that he has to get answered — and he hasn't had much practice and he doesn't have much skill in how to answer these questions. Right off the bat, he wants to know what kind of work will be appreciated in this place: "Will they want me to do what I want to? How do I find out? And to what extent do I believe the words I hear?" Second, he asks himself, "Will I be *capable* of asking major productive questions? My thesis adviser posed my thesis question for me. I have no experience *asking* technical questions!" Third, he asks: "What is this thing called a boss? He is not a thesis adviser, or a senior professor. I understand *those* kinds of people, because I've been around them for several years. But I do *not* understand 'boss.' How should I behave toward him? What is the extent of his influence over me? Is this a legitimate extent? What are the mechanisms that he has to influence my behavior? How do I find out about them? How much should I *take* from him? What happens if I do or I don't?" The young man must answer his questions via a difficult learning process. It is a *long* process, and throughout it he suffers anxiety and uncertainty. Fourth, he asks: "How do you talk around this place? What is the attitude toward working habits? What are the do's and don't's of this organization? How much identification is the organization going to expect me to have with it? Will I find that amount intolerable? Will my degree of identification here be compatible with the indentifications I have — and wish to establish — with other organizations?" And finally, he asks: "Am I going to achieve my personal goals here? Whatever it is I want to do of myself — and maybe I'm not sure yet what that is — am I going to be able to do it here?"

These are questions that trouble the young people in the *best* industrial laboratories. When we move down the scale, to laboratories whose environments are not so conducive to good science, we encounter some of these same questions plus others: "Will they permit me to work on this? Will I be in trouble if it fails? Will they want it if I *succeed?*"

What to do about such problems? For one thing, management simply has to learn what scientists are like; then the precepts and practices of management have to be modified to accommodate those modes of scientist behavior that exist among effective research people. But this is only a partial answer, for it says nothing about the young scientist's preparation for industrial research. He comes *un*prepared, at least psychologically, because the people he knew in graduate school were not of this industrial world at all. Rather, they were people like Weis and Walker — people who saw no relevance in words like "relevance." Moreover, the problems he worked on as a graduate student were not problems that *he* had created. Most likely they were problems his mentor had given him, problems in the faculty man's sphere of interest. Says Bell Labs' Jack Morton: "The young scientist has taken more direction from his thesis adviser than he will *ever* get from his boss — *if* the boss is good and if he understands relevance." And, as Steele points out, when two out of five graduates end up in a career that is different — by anything from a little bit to a lot — from the career they were prepared for, then a serious mismatch exists. He is cautious in suggesting a solution here: "Graduate education is clearly far too important to tinker with casually, but by the same token it is also too important to ignore."

One hears only an occasional cry that graduate education in the sciences be overhauled and tailored more exactly to the needs of industrial science, but what is proposed occasionally is a diagnosis of the problem that may be almost as dangerous. This is the notion that the problem relates to the corporate "image," and that many graduate students either refuse to consider careers in industrial science, or enter such careers with reluctance, because they have a "bad image" of the industrial world. This diagnosis is dangerous, for it suggests that cosmetic solutions will solve a very serious problem.

The nature of that problem, of course, is simply that we know very little about running these enterprises. Indeed, the more one tries to *run* them, the more stubbornly they refuse to be run. This is a costly kind of ignorance, as we see in the woeful tale I heard not long ago from a scientist with intimate knowledge of a very large industrial laboratory in one of the world's largest industries. Some years ago, there began to grow a cleavage of suspicion between the laboratory and top management of the company. At first, it was a matter of the laboratory's failing to create the kinds of things top management was asking for.

When the laboratory itself proposed a new thing, then carried it through development, the thing would more often than not fail to establish itself in a market. Usually, this was not a fault of the product, but of management's inability in the marketplace. Still, such failures caused a fear on the part of management toward *all* developments that the people of the laboratory proposed — a fear that these too would fail in the market. And finally, the ideas stopped flowing altogether. This is the situation today, though the laboratory continues to cost the company about one million dollars per *week!* What if this laboratory ceased to exist? What effect would this have on the company's competitive position? The man who told me the story seems to think that, aside from the technical services it provides, it would have no effect at all!

We might put several labels on this story, for it illustrates "communication breakdown," "dictation," "loss of trust," and perhaps one or two other frailties on the part of both parties. But one point that is not suggested by this particular example — and yet a very important element in our understanding of the problem — is a point that might be labeled "motivation." In particular, it is the matter of the *distinct difference* between the motivation of the top manager and that of the scientist.

Certainly one very powerful motivation of the top manager, if not his prime motivation, is to make a profit. He is measured by his ability to do this and, indeed, he is rewarded accordingly. In turn — and perhaps instinctively — he measures the worth of the industrial scientist by the same rule: Is he profitable or is he not? The deadly thing about this question is not the threat it poses to the "not profitable" scientist. Rather, its deadliness is the fact that few scientists — however creative they may be, however "profitable" they may be — find sufficient satisfaction in their profitability. To know that his discovery is reaping great profits for his company may make the scientist feel good as an *employee*. Indeed, he wants to know that he is paying his way, that he is valuable to his organization. But knowledge of profit does not satisfy him as a *scientist*, even though he may share in the profit. The scientist's reward is knowledge that he is making a contribution to science. This is the goal that every scientist seeks. If he is a member of an industrial organization, his objective is likely to extend even further — to the hope that his discovery will lead to some social good. These are the things that motivate the industrial scientist — "Am I contributing something to science? Am I contributing something to society?" Profit is an insufficient measure of either.

David Allison

2

Creativity

A manager of design engineering, who is a friend of mine, brought me the following problem. He supervises some 18 design engineers in a company that manufactures consumer appliances. Because the company had recently suffered a severe financial setback, everybody was under strong pressure to cut costs and turn out a competitive yet inexpensive product. The manager was worried that, with all the emphasis on the current product, his men were neglecting to think about future markets and future products: Over the past year, they had not turned in a single patent application.

T.A. EDISON ELECTRIC LAMP

PATENT No. 223,898 1/27/80

My friend realized his company had to think of future markets as well as present ones. He believed his group had to turn out patents in areas that might be important in tomorrow's marketplace if his company was to survive. To get more patents, he reasoned, he had to increase the creativity of his group. So the problem he posed to me was how to go about increasing creativity.

His first impulse was to replace some of his men with other engineers who would be more creative. For this purpose, he asked for a test of creativity that would help him to select such creative engineers. After I discouraged him from this, his second impulse was to re-educate his men. He thought of instituting a course in creative engineering. But he vetoed this idea himself after considering the cost and time involved. His third impulse was to find some incentive to motivate his men to be creative. He thought of a contest or some sort of bonus plan.

It was in connection with this alternative that he asked my advice. He wondered if the idea was too "gimmicky." I asked him what his men thought of the idea. "I'm embarrassed to tell you," he said, "that I have never discussed the problem with them."

On the same day, he called his men together and told them the problem: "We need patents or we won't survive." Within a month, several patent applications came from his group. And during the next month, the number increased again.

This story illustrates some of the ways management seeks help from research in creativity. Management — and other potential consumers of creativity research — expects psychologists to provide tools for selecting, training, or motivating men to create. In response to such expectations, and because of financial support for such purposes, the bulk of the research in creativity is explicitly aimed at providing such tools. How good are these tools?

To help identify and select creative personnel, many investigators seek to find what sort of people are more creative. The Institute of Personality Assessment and Research at the University of California, for example, has devoted the past several years to a search for the distinguishing personality traits of artists, writers, architects, mathematicians, and engineers. In the study of architects, the psychologists chose 40 of the most creative people in the profession, using the nominations of professors of architecture and editors of architectural magazines to help select the group. These creative architects spent three days on the Berkeley campus, where they were observed, interviewed, rated, and tested for a number of personality traits, an assessment procedure that yields some 1,000 measures or indices for each architect. The psychologists then sought those indices, alone or in combination, that differentiated creative architects from less creative architects, from "average" individuals, from creative groups in other professions, etc. Creative architects, for example, seem to have more theoretical and aesthetic interests, higher self-confidence, greater preference for asymmetrical and complex design than do their less creative counterparts. Thus, the Berkeley psychologists focus on creativity as a personality and temperamental characteristic.

Other psychologists treat creativity as an intellectual quality or ability. Guilford's research at the University of Southern California, for example, concentrates almost exclusively on sorting out creative abilities in terms of performance on short, paper-and-pencil tests. A typical test has a time limit of about ten minutes. In any one study, Guilford and his colleagues will administer approximately 50 different tests to several hundred men, usually a captive audience of Air Force or Navy personnel. Some tests deliberately tap well-known abilities such as verbal fluency, vocabulary, arithmetic facility, deductive reasoning, and the like. Others are deliberately constructed to tap creative abilities other than those that may be tapped by existing tests: One describes the plots of a few stories and the subject is asked to invent clever titles for these stories; another, aimed at tapping originality, requests the subject to invent captions for cartoons. To get at "sensitivity to problems," one test instructs the respondent to suggest what is wrong with existing social institutions; another asks him to suggest improvements for common household appliances. In still another, the testee is asked to list as many uses as he can for a common red brick.

Guilford tries to discover the minimum number of independent abilities that underlie performance on these various tests. He finds, for example, that performance on the cartoon captions and invention of clever plot titles seem to go together. Because the scores on these two tests are independent of scores on tests of known abilities, Guilford infers that he has isolated a new factor. In this particular case, he names that factor "originality." But he has yet to demonstrate that his creativity tests are related in any demonstrable manner to creative performance in the real world. Nevertheless, his work has become the basis for almost all current attempts to construct an applied test of creativity. *The AC (Sparkplug) Test of Creative Ability*, for example, reflects Guilford's categories: The subject thinks of reasons to account for a given fact, lists possible consequences of a particular happening, suggests other uses for a common object, and indicates limitations and improvements for household appliances.

So far, none of these tests for creativity has met the requirements for practical use, such as those listed by the American Psychological Association. Indeed, until one can specify quite explicitly the quality he wants to predict, it is almost meaningless to discuss whether a test for creativity — or any test — can be of possible use to him.

Other investigations concentrate on training individuals to be creative. They search for conditions or training procedures that facilitate creative achievement. Or they try to evaluate the effectiveness of existing training programs. Unlike the work in creativity testing, these investigations start with the assumption that creativity can be changed by imparting skills, knowledge, and techniques during a training program.

Research in creativity training uses principles based on free association. Two interrelated themes guide such work. The first is that creativity is facilitated when the thinker suspends judgment while he is generating his ideas. The second is that quantity of ideas will lead to quality of final solution. One source of these principles is psychoanalytical theory and practice, where emphasis is placed on "releasing" the individual from debilitating constraints. Another popular source is advertising executive Alex Osborn, who has incorporated these principles into his famous "brainstorming" technique. Sidney Parnes and his colleagues at the University of Buffalo, supported by money from Osborn, have conducted experiments that they feel support the quantity-quality hypothesis. They ask individuals to think up uses for a common object — say, a newspaper; the individuals are told to "free associate," to generate as many uses as they can, good or bad. Then these ideas are compared with ideas produced by other groups — groups who were told to give only their "good" ideas. Parnes finds that brainstorming yields more "good" responses. He has also studied the effectiveness of a course in creative problem-solving. He finds that students who take the course get better scores on a creativity test than do students who did not take the course. Such research, however, is far from conclusive, as Parnes himself will admit.

Other studies, including ones I've done, raise serious questions about the existence of any simple relationship between quantity of ideas and the quality of the final product: Quantity of ideas can facilitate, hinder, or make no difference to the quality of the final solution. It all depends on a number of other factors: the nature of the task, the prior training of the thinker, the stage in the problem-solving process, and the content of the pool of associations from which the thinker is drawing.

The evidence for the effectiveness of the creative problem-solving course must similarly be qualified. For one thing, other experimenters, including myself, have gotten similar increments in test scores by merely instructing subjects to "be creative," rather than having them go through a full training program. For another thing, we have no evidence that increments in test scores bear any relationship to performance outside the class or testing situation.

So far, research in creativity has provided only indirect evidence about the effectiveness of incentives and rewards upon creativity. The work of Pelz at the University of Michigan suggests that scientific productivity tends to be correlated with a value system that orients the scientist to his profession rather than to the particular institution for which he may be working. Other work, relevant to environment and leadership, relates to the effect of group composition on creative performance: Hoffman and Maier at the University of Michigan deliberately compose groups of students that may be homogeneous or heterogeneous with respect to personality traits or sex. These groups are assigned the task of solving problems — usually of a human relations type — by means of role-playing. The investigators find, in line with their hypothesis, that the more heterogeneous groups, when they *do* reach a solution, usually produce a more "creative" solution than do groups who tend to "see things in the same way." (I will have more to say on this in a later chapter.)

What is the effect of environment and leadership on creative performance? I find it significant that my manager of design engineering — mentioned earlier — thought only of changing the men under him as a way to increase creativity. He did not ask about replacing or retraining himself. Nor did he suggest changing the work environment, the compensation system, the "climate" for creativity, or otherwise altering the company situation to fit the workers, rather than vice versa. Managers often turn to psychologists for tools to implement their objectives. They want research in creativity to give them the means to select or change men to function within the existing framework.

I believe such hopes are shortsighted, for they are based on a misconception of the most useful role of psychological research in the area of creativity. The manager's quest for tools to help him implement existing policies will yield a poor return for the investment made in research aimed at providing such tools. Let me go on to say why.

Let's assume the unlikely possiblity that psychologists can succeed completely in supplying management with tools to select and to train men to create within existing setups. Such "success" may create some immediate, short-run gains, but these gains might come at the expense of long-run adaptability and flexibility. The success of tests or training programs can only be measured in terms of current standards and values. To the extent that we maximize the adaptation of men to a particular set of standards, we may irretrievably lose the necessary flexibility to survive when these standards and values change in response to changing values, consumer demands, and economic situations.

Even more serious is the possibility that the use of such tools may backfire in terms of increased frustration and resentment on the part of employees. Unless we simultaneously overhaul the entire company from top to bottom, the successful selection or training of men for creativity can produce false expectations. What can creative men accomplish when they enter an environment that is not prepared for them, an environment that may be hostile, that fails to provide opportunities that such individuals need?

The following incident will illustrate this last point. I was interviewing some managers concerning their views on hiring engineers who had been through a creative engineering program. When one manager told me that he had hired several engineers from such a program, I asked how these graduates compared with his other men. "They're good," he explained. "The only complaint I have is that it takes me two years to beat all the creativity nonsense out of them. Then they settle down to do what they're told to do."

Before I turn to my own recommendations on what to do about creativity, it may help to pause for a moment to consider what we mean by it. Because "creativity" is a noun and because psychologists study "creativity," we sometimes carelessly talk as if it is a "thing" to be studied in its own right, something we can isolate and deal with as a separate entity. But it is not a thing or entity; rather, it is a *quality* of things. To some, creativity is a quality of people; to others it is a quality of behavior or performance; and to still others — the manager of design engineering, for example — it is a quality of products.

The previous discussion about what managers expect from research in creativity and the description of what psychologists are doing under the name "creativity research" imply these different meanings. The work on personality assessment and Guilford's investigations deal with creativity as a quality of people. It represents another attribute by which we can sort people into groups or by which we can order them along a continuum. For one group of psychologists, "creativity" is a quality much like temperamental and personality traits; it represents a tendency to feel, value, and act in certain ways. For others, such as Guilford, it is a quality more like intelligence, representing a capacity to perform with certain skills or at a given level.

Research that is oriented toward training deals with creativity as a quality of performance, or of different states of a particular individual. It represents another way of differentiating one state of an individual from a previous state. Thus it makes sense from this viewpoint to talk about a particular individual as being more creative on occasion A than on occasion B.

Much of the interest in creativity, especially by industry,

does not refer to a quality of individuals or performance; instead, the concern is with "creativity" as a quality of products. There is nothing wrong with this particular use of the term "creativity," so long as we keep clear the distinction between the creativity of the *outcome* of a man's efforts and his creativity in *producing* that outcome. In turn, these usages of the term "creativity" must be distinguished from the more general, relatively stable "creativity" characteristics that set one man apart from others. In light of these different meanings, it makes sense to refer to the painting of a chimpanzee as "creative" while, at the same time, denying that the chimpanzee is creative or that he behaved creatively. Also, it makes sense to deny that the solution to a well-known puzzle is creative, but to describe a child's behavior in achieving this solution as creative because he had to perform in ways that were novel in terms of his own past experience. And finally, we know the meaning when someone laments that so-and-so — a creative individual — is wasting his time in uncreative work. These different meanings and expectations can cause misunderstanding, especially between psychologist and client. The psychology of creativity must restrict itself to creativity of performance and/or of individuals; the relationship of such creativity to creativeness of products may not be close.

One implication we can derive from this discussion of the meaning of "creativity" is this: While industry and management focus purely upon the results or payoff in terms of the creativity of products, the psychology of creativity focuses entirely on people and their behavior. To a manager, it makes sense to pay for results — to reward a man only after his contribution has been judged to be creative. To a psychologist, such a compensation system makes no sense at all, because "creativity" that refers to results cannot be treated as a meaningful, functional entity in psychological terms. Hence, the psychologist will tell you: To encourage creativity, you should reward a man for behaving creatively, even when he fails entirely. Because the factors that lead to a "creative" product are complex — and often accidental — rewarding for the product may actually reinforce accidental or noncreative aspects of behavior. Such a reward system may lead to superstition and false notions. It may, alas, even stifle creativity. Emphasis on the "creative" product, like emphasis on the grade in school, may lead to a mistaken set of values, to imitation rather than to innovation.

If the attempt to provide tools to help management develop creativity within the present industrial framework is shortsighted, what positive steps can we take? One answer, of course, is implied in my earlier criticism of the one-sided attempt to change employees without changing the environment into which they are placed.

It is at the level of company policy, management objectives, and organizational principles that I believe psychological research will have its biggest impact, if it is to have any impact at all. To have an impact at this level we must think of creativity research in terms other than application and tool orientation. I believe the type of creativity research that will provide this impact is that which we call "basic" or "pure." I use the term *basic* to mean research that has *no* applied objectives, that is not oriented toward payoff, either now or in the long-range future. In this sense, basic research refers to investigations of creativity whose major objective is an understanding of the most general laws that govern man's psychological behavior in the realm of creative achievement, and whose only justification is in terms of theoretical significance or potential ability to help us clarify issues at a fundamental, conceptual level. The goal is not payoff; the goal is purely one of greater understanding.

Basic research, then, does not aim at producing tools or prescriptions. Rather, it aims at helping us make explicit our basic assumptions about man as a creator. It does this to enable us to test these assumptions, revise or correct them, and eventually replace our present image with a more adequate one.

It is at the level of our assumptions and our image of man, rather than at the level of test and training principles, that psychological research may possibly influence company policies and objectives: When we examine the basis of current policies and organizational principles, we find each of them based upon some bedrock of major, often implicit, management assumptions about human nature. Such assumptions, because they are implicit and rarely brought to the scrutiny of consciousness, often contradict one another. One management assumption, for example, might be that employees are "rational" and will choose to vote in favor of the company if they are presented with all the "facts." Another assumption, often made by the same policy-makers, is that workers are like children and do not know what is best for themselves; therefore management should make decisions for them about retirement plans, stock bonus plans, etc. Another assumption is that men are guided by the profit motive, that without adequate financial incentives, men will not work their best, etc.

Such assumptions do not hold up well under psychological investigation. The areas management has reserved to itself — union policy, compensation principles, organizational levels, division of responsibility, decentralization or centralization, etc. — imply major assumptions about what makes men behave as they do. The hope for creating change at this level lies in basic reasearch that goes right to the heart of the assumptions we tend to make implicitly about ourselves and our fellow humans.

Such changes in our image of man do not come about easily or quickly. But they do come about, and they can be brought about. Witness the great changes in our image of ourselves that came because of the theoretical and scientific speculations of Darwin, Pavlov, and Freud. Freud, alone, has been responsible for widespread changes in how we deal with individuals — not only in psychiatric practice, but also in political and economic policy, in legal decisions (even in the Supreme Court decision on segregation). It was Freud more than any other man who established the image of man as the slave of powerful, often unconscious, motivations and irrational determinants.

In our own times, the man who seems destined to have an impact upon psychology and on man's image of himself, comparable in scope to that of Freud and Pavlov, is the Swiss psychologist, Jean Piaget. For almost 40 years, most of which he was largely ignored by the mainstream of psychology, Piaget has studied the development of thought structures in children from birth through adolescence. It is only recently that American psychologists have been aware of the enormous implications that Piaget's work has for our understanding of the nature of intelligence and intelligence tests. Until the Second World War, American psychology and the American school system were dominated by the assumption of fixed intelligence and fixed development of intellect. Some assumptions led to notions about child rearing and education that stressed, for example, the desirability of keeping the child from being overstimulated during his formative years, lest he be overloaded or challenged. Partly as a result of Piaget's fundamental work, we now believe that an individual's intellectual development depends crucially on the amount and kind of stimulation he had during his formative years. The implications for child rearing are now almost completely the reverse of what they were with the older assumptions. Instead of protecting the child from too much stimulation, the implication is that, if he is not stimulated, the child will fail to reach his full intellectual capacity.

What do we conclude from all of this? Especially, how can creativity research relate to current management practice? I believe that psychological investigations into the nature of creativity can have a major impact on industrial and management practices. But . . .

In seeking tools and prescriptions for increasing creativity, management is asking the wrong questions of psychological research in creativity. Such research is limited in its scope and generality. At best, its results are transitory. Even when such research succeeds, the ultimate benefit may be negligible or even negative. This is because "success" has to be measured in terms of present standards and objectives . . . and these standards and objectives may be the barriers to creativity within any company.

Although management tends to deal with the "creativity problem" by seeking ways to change its employees, any success in changing the employees' creativity may boomerang, unless — at the same time — changes are made in the company organization and leadership. False expectations and consequent resentment can occur when creative people are put into an environment that is not ready for them. Hence, before changing the environment *and* the creativity of the people in it, we must modify the assumptions on which the environment is based — company policies, unwritten rules, etc. Such assumptions can only be altered by basic research in creativity: research whose objective is *understanding*.

Hence, though I believe psychological investigations into the nature of creativity *can* influence management practice, I believe the major impact of such research will come at the level of our basic assumptions about the nature of man. And this new image of man can come only from fundamental research in creativity. Research that aims at giving managers what they now think they want — better tools to implement their current image — will contribute little.

Ray Hyman

3 Blocks to Creativity

The attempt to be creative in any field makes extraordinary demands on the human personality. In science and engineering, creativity demands a flexibility that may be greater than that of almost any other occupation to which man can dedicate himself. The scientist and the engineer must first of all maintain mastery of an enormous body of rapidly growing data, yet, at the same time, must be as freely imaginative as the poet, the artist, or the musician. He must also possess the capacity to anchor his imaginative flights to reality, to test the degree to which they are consonant with reality, and finally to project them into the future for new uses. Few other occupations demand so much.

I do not deplore the fact that creativity in science makes this demand on us, but only that we do not recognize its implications for the education of the scientist. I deplore the fact that we do so little to help him attain — as a student, and then as a matured scientist — that degree of emotional stability and freedom that is essential if he is to use his intellectual endowments in the most creative and constructive way possible. I am unhappy at our complacency with a primitive educational process that re-enforces many of the concealed but universal neurotic ingredients in human nature.

This concern leads to two questions: What is it in the educational process that limits and inhibits creativity? And what can be done to change the educational process so as to lessen its damping action on potential creativity? Even before discussing this, however, we must be clear about the nature of the creative process and its inherent vulnerabilities. This can best be approached through a few examples.

There are scientists and engineers who produce brilliantly as students, and even through their graduate years of study, but then collapse. There are gifted investigators who turn out one or two creative achievements early in their careers, but thereafter are never productive again. Others seem to acquire creative freedom only slowly and relatively late in life, and therefore fail ever to win security, tenure, or recognition. Others are consistently and repeatedly successful and productive, but notwithstanding become increasingly despondent and dissatisfied. Some young scientists become blocked after marriage, while for others the reverse is true. Many creative minds go through long periods in which they are barren. For some, these prove to be enriching periods of lying fallow. For others, they are years from which their creative talents never recover. For Mark Twain and for Beethoven such periods were nearly catastrophic.

An early work block may impede a young man's development for years or may destroy him, so that he may never be able to return to his profession. Later in a scientific career, it can turn a promising future into a wasteland. A still later work block may stop a mature scientist so completely that he may have to save himself from scientific oblivion, and his family from want, by turning to teaching, administration, work for a foundation, or even a college presidency.

The outstanding figure in a certain field became scientifically sterile after he had become a full professor and department head. This sterility so embittered him that, although he was a passionately dedicated teacher, he could never allow any student to become creative, to publish, or even to escape from him without a fight. Furthermore, he recommended for important posts only those who were shaped in his own image and were similarly sterile.

How do such work blocks arise? We know from clinical studies of creative individuals from the sciences, the arts, and engineering, that there is no single cause, just as there is no single cause for fever. But we also know that the creative process in each of us is vulnerable to distortion and inhibition by the neurotic process, re-enforced by neurotogenic ingredients in the educational process.

What, then, is meant by the neurotic process? I will not contrast a hypothetical neurotic *man* with an equally hypothetical normal *man*. I will try instead to characterize the neurotic process by pointing out the essential difference between a normal and a neurotic *act*. It is important to keep this circumscribed purpose in mind. We must first understand what constitutes the essential disorder of a single act before attempting to characterize a total personality as sick or well.

Everything a human being does can be either sick or well, neurotic or normal. Whether it is feeling, thinking, eating, sleeping, drinking, fighting, killing, hating, loving, grieving, exulting, working, playing, painting, planning, inventing — it can be one or the other. Further, the category to which an act belongs does not depend upon conformity to any cultural norm, nor to whether the act or feeling or thought is intrinsically sensible or foolish, useful or valueless, constructive or destructive. The measure of health is dependent upon none of these criteria. The measure of health is flexibility, the freedom to learn through experience, the freedom to change with changing internal and external circumstances, to be influenced by reasonable argument or by the appeal to emotions, and especially the freedom to cease when sated.

The essence of normality is flexibility in all of these vital ways. The essence of illness is the freezing of behavior into unaltering, repetitive, and insatiable patterns. We see examples of such frozen behavior in all creative fields. In painting, we see it in men of worldwide reputations, men who, after passing through some inner convulsion of the spirit, start on a new "period," dominated perhaps by a new color, or new subject matter, or a new way of applying the paint, or a new way of stressing outlines, or a new way of distorting proportions, each such innovation soon becoming as rigid as the work of the earlier periods.

Patterns of scientific creativity provide examples of comparable distortions. There is the man whose anxieties compel him always to rush, always to be ahead of time, whose work must always be done long before it is necessary. In his hurry, such a man can barely stop to reflect, to read, to digest, to lie fallow. His opposite number gets to the train only as it is pulling out of the station and can never think about a problem until he is under last-moment pressure. He cannot lie fallow either. In both, the neurotic patterns of behavior exercise distorting influences on the processes of study, of imaginative rumination, of creative experimentation. It cripples the capacity to assemble new data or to arrange old data in new patterns. I think of three scientists of high potential in three different fields whose creative capabilities were distorted by destructive compulsions always and only to criticize. One of the unique honors of science is the humility and honesty with which it tries constantly to correct its own errors. Healthy criticism — of self and of others — is good. But these three men misspent their lives creating nothing, because obsessional mechanisms, driven by unconscious hate and envy, took over the intrinsically useful function of criticism. They did nothing but criticize: today themselves, tomorrow others. The seesaw never ceased.

Another example was a brilliant pharmacologist who tried to establish a hypothesis about drugs and drug action; but his very hypothesis was chosen unwittingly under the influence of concealed neurotic mechanisms, and when he set out to prove it, these same neurotic influences determined the design of his experiments and the techniques he employed. The work on water metabolism of another scientist was skewed by his years of struggle with bed-wetting in childhood, and the humiliations to which a scornful father had exposed him.

What do such observations say to the old cliché that one must be sick to be creative? Some psychotherapists hold this view, as do many laymen; yet this culturally noxious assumption is devoid of truth. No culture known to us has succeeded in bringing up children and adults free from concealed neurotic mechanisms. Therefore we have never known creative folk in science or the arts who were exempt. The creative capacity of a few has surmounted this obstacle. But it is a fallacious non sequitur to conclude that without the obstacle they could not have been creative.

The creative potential in each of us is *not* dependent upon nor is it derived from the neurotic potential. But the creative process can be distorted by neurotic mechanisms. These arise in early childhood, not out of exceptional circumstances, but out of simple and ubiquitous human experiences. With age, the conflict between the two processes usually is intensified by later stresses, one of which we euphemistically call "education." And when any man succeeds in being creative, he does this *in spite* of this neurotogenic conflict. Indeed, the conflict between his creative process and his neurotic process causes his actual creative productivity to fall far short of his potential productivity. That fragment of his potential creativity that survives the impact of the neurotic process is distorted in content and becomes rigid and stereotyped.

Yet there are many complex reasons why the creative person clings to the myth that his neurosis is the source of his creative capacity. Instead of feeling proudly that his creativity is that in him which is unique, he makes light of his extraordinary gift in favor of the rigid, banal, undistinguishing component of his nature, i.e., his concealed neurosis.

I said earlier that the essence of neurotic illness is the freezing of behavior into unalterable and insatiable patterns. This happens whenever a configuration of psychological processes

The man who is trapped: His preconscious processes cannot function freely, because unconscious processes are dominating his activity. Without realizing it, he will use his special competence and knowledge to express the conflicts of his own spirit. The man may perform creatively, but he does so in spite of these neurotic mechanisms. If he could free himself, he would increase his creative productivity significantly.

predetermines the automatic repetition of behavior. How this occurs is a complex story.

There is evidence that our psychological processes represent the interaction of concurrent systems. We call them concious, preconscious, and unconscious. About their continuous interplay, we know that man behaves with greatest flexibility and freedom, i.e., most normally, when the preponderant influence is exercised by an alliance of his conscious and preconscious processes. And we know that he operates in the shadow of illness whenever unconscious determinants are dominant. These may strike you as convenient theories rather than as factual, yet they rest firmly on a large array of experimental and clinical data.

In recent years our concepts of the interplay between conscious, preconscious, and unconscious processes in human psychology have changed in a direction that will lead to more precise clinical and experimental work. We now look upon the brain not as a device to do work but as a communications machine to transmit information. At the core of this process is a continuous stream of subliminal, i.e., "preconscious," activity, which goes on both during sleep and when we are awake and is carried on without conscious symbolic imagery. Analogous to a computer, it processes "bits" of information by scanning, ordering, selecting and rejecting, arranging in sequences, by juxtapositions and separations on the basis of chronology, by condensations on the basis of similarity, dissimilarity and contrasts, proximity and distance, and finally summating and coding.

This preconscious processing of data proceeds at an extraordinarily rapid rate and with great freedom, as it assembles and disassembles many diverse patterns. The process is both analogic and digital. Let me repeat that none of this preparation of data is conscious, yet this is what implements the processes of thought.

Furthermore, the stream is fed by a continuous bombardment of messages that are signalled from changes in every aspect of the body's functions. It is fed also by an incessant bombardment of signals from changes in the outer world that reach us through our distance receptors. Therefore, we can divide this input into that which comes from the surface of the body — i.e., the bones, muscles, joints, tendons, skin, and apertures — the input from the internal organs, and the input from a distance. Like the central processing itself, the major part of this incessant bombardment is subliminal. Our highly developed symbolic processes *sample* this entire stream of preconscious activity, i.e., the input, the internal processing, and the output. This *sampling* process is conscious; and it is this conscious sampling of the stream of preconscious input, preconscious central processing, and preconscious output that is always miscalled "thinking." This process of conscious symbolic sampling has an important function in mentation; but it is not thinking. Rather, its function is to relate the samples of the preconscious stream to reality, to test them, to ruminate about them, and to communicate them to others. In turn, all of this is fed back into the preconscious stream itself, where it exercises a continuous cybernetic influence. Note, again, that learning, thinking, and creating are all preconscious, while this process of sampling, ruminating, testing, and communicating is conscious.

How, then, do unconscious processes enter this picture? They arise out of distortions of the relationship of conscious symbols to their roots. The symbol itself is never unconscious; it is the link between the symbol and what it is supposed to represent, which becomes distorted, displaced, unconscious. This distortion in turn feeds back many secondary disturbances and is an essential ingredient (perhaps *the* essential ingredient) in the neurotic distortion of human nature.

There are many ways of formulating these relationships; but this seems to me to be the one that is most relevant to the problems that confront us here, i.e., the problems of learning and, above all, *the problem of how to protect the freedom of preconscious processing in learning, in education in general, and in creativity*. The freedom sought here is freedom from the destructive effects of an "educational" input overload; freedom from excessive anchoring to literal, pedestrian, conscious realities; and freedom from neurotic distortions. This is the great challenge that scientific education (and indeed all education) has never faced in the past but must face now.

Both conscious and unconscious processes are fixed and rigid. Normally, the conscious symbol is anchored to literal reality. Conversely, the symbol whose relationships have become distorted and unknown is bound even more rigidly to an unknown. The symbol here is to its hidden root as a delegate who has been sent to a conference table to "negotiate" but with secret orders never to modify his position. We see this stereotyping influence in the works of creative people in many fields — the man who paints the same painting over and over again; the poet who writes and rewrites the same poem; the scientist who grinds the same scientific ax. It is this that leaves its personal signature on a painting or musical composition and accounts for the man who produces only one play, one book, one piece of first-rate scientific work. This is the price we pay whenever unconscious processes hold the upper hand.

How, then, do creative processes operate between these two rigid systems? This depends upon the freedom of preconscious functions, which are the implements of all thinking, particularly of creative thought. Preconsciously, we process many things at a time. By processes of free associations, we take ideas and approximate realities apart and make swift condensations of their multiple allegorical and emotional import. Preconscious processes make free use of analogy and allegory, superimposing dissimilar ingredients into new perceptual and conceptual patterns, thus reshuffling experience to achieve that extraordinary degree of condensation without which creativity in any field would be impossible.

We must remember always that all three processes act concurrently. Whatever we do, whatever we say, whatever we think — whether we are sick or well — everything is a composite resultant of all three processes. This is true when we dream, when we are creative, or when we are preoccupied with the banal events and affairs of life. The differences are in the relative roles of these components. When a scientist is studying atomic energy or a biological process, when an engineer designs a bridge or a spacecraft, when a classicist studies an ancient language, each deals with his subject on all three levels at once.

The man who is free: His brain is functioning as a communications machine, processing bits of information by scanning, ordering, selecting, etc. This is preconscious processing and it proceeds at extraordinary speed. The symbols shown on the screen represent a sampling of this preconscious activity. Here, our man is relating this sample to reality: His preconscious provides him with myriad bits; he samples these (a conscious activity), tests them, then back they go into the preconscious.

On the conscious level, he deals with his subject in terms of communicable literal ideas and approximate realities. On the preconscious level, he deals with swift condensations of their multiple allegorical and emotional import, both direct and indirect. On the unconscious level, without realizing it, he uses his special competence and knowledge to express the conflict-laden, confused, and hidden levels of his own spirit; and to the extent to which unconscious processes dominate his activity, he will use the language of his specialty as the vehicle for the outward projection of his internal struggles.

We saw this in the examples cited earlier: the men who did nothing but criticize; the pharmacologist whose hypotheses were chosen under neurotic influences; the scientist whose work on water metabolism was distorted by childhood problems with enuresis. In each of these the distortions were happening without their knowledge, precisely because unconscious processes had taken over. Each man's creative thinking was being subverted to serve his unconscious needs.

Not only are the products of the preconscious thought stream vulnerable to distortion from unconscious levels, the stream itself must be protected from the same influences, because creativity depends upon its free flow. Let me cite a few examples to re-enforce this point: There is the famous reverie by Kekulé of the six snakes in a ring, each swallowing the tail of the one ahead, an image that illuminated for him the structure of the benzene ring. There is Otto Loewe's account of the derivation of the whole neurohumoral concept from a dream. There is Poincaré's solution of a mathematical problem during a state of reverie on a bus.

Such experiences indicate that new syntheses occur when preconscious processes can operate without the restrictions of conscious processes and without interference from unconscious determinants. I deliberately plan my work to capitalize on it. I usually work on a problem until I am tired late in the evening. The moment my head touches the pillow I fall asleep with the problem still unsolved. Frequently I will awaken two or three hours later — sometimes to find myself

The man with an idea: Here he has thought (in the preconscious) and has tested the new formulation (on the conscious level). If his formulation passes the test of reality, he may then communicate it (also on the conscious level).

in the middle of the very sentence on which I was hung up as I went to sleep — but my head now filled with a rushing tumult of ideas and phrases, a new assembly of the material. I have learned by sad experience that I will have forgotten most of this by the next morning unless I dictate or write a note about it *at once.*

Creativity in technical fields puts many special demands on the individual. In the exploratory phase, i.e., while gathering data, one's observations should be free from preconceptions and preferential biases — and, above all, free from any drive to systematize the data prematurely or to formulate hypotheses prematurely. The work of this exploratory phase requires a free, imaginative, flexible, uncommitted attitude of fact gathering, i.e., the attitude in which preconscious functions predominate.

Later steps of scientific or technological investigation demand that this material be subjected to more rigidly organized psychological processes, which enable one to test and then to doubt, then to test and doubt again, and then to rearrange the data into new systems. This requires a personality that can doubt on a consciously self-critical level, yet that will not be caught up in obsessional indecision.

The work of the man who becomes trapped may suffer in various ways. One man cannot rest until a pseudosolution is in his hand. Like the kleptomaniac, he may be forced by phobic and unreal anxieties to attempt to increase his feelings of certainty by falsifying data even after fully convincing evidence is already in hand. Another is afraid to find answers, his terror increasing as he approaches any solution. This forces him to postpone endlessly any definitive conclusion of his work. Or he chooses problems that will take a lifetime to solve. Or he clouds his findings in obscure terminology. Or he refrains from publication altogether.

Clearly, creativity in science and engineering demands flexibility and imaginativeness, but also tightly organized thought processes, matched by a high degree of emotional and psychological freedom. In science, the moment of apparent illumination is never the end of a search. It is always the starting point for new investigation. It provides the scientist with a beckoning goal and an incentive, without which nothing new is ever discovered.

Thus the impact of unconscious conflicts on the creative process is strong. If conflicts are buried under distorting disguises, they block the flow of preconscious analogic processes, which are essential to scientific creativity. If, on the other hand, a scientist understands his conflicts, if they are out in the open, he can use them. They may spur him on. They become a supplement to, but not a substitute for, his fundamental biogenetic and instinctual processes. He guides them, instead of being enslaved by them. The man who lacks such insights is enslaved by his unconscious conflicts and drives. He lives on a treadmill, for he can never finish one piece of research without being seized with terror that he will never be able to complete another. He can never lie fallow, never rest. Without work he is in a torment of anxiety; and complex compulsions take over his normal creativity. "Success" always cheats him, because success never enables him to attain the unconscious goals of his activity. To him it is not the need to solve his chosen problem that drives him, but an unknown necessity. His consciously chosen goal is merely a substitute for these unconscious and unattainable goals. Therefore, even when he reaches some consciously chosen goal, he feels defeated. He may achieve success by many standards, but he can never enjoy the achievement. Unless he resolves his inner conflicts, he can never get off the treadmill. In his middle years such a man is likely to become angry and depressed. Indeed, even in spite of any degree of careeristic success, the likelihood of middle-year depressions is greater with him than with others among his peers who were less "successful." And a "nervous breakdown" in such cases is not a result of overwork, but a consequence of a life driven blindly and without sufficient insights.

I spoke sharply of our "primitive" educational practices. I shall make only a few tentative suggestions as to how the educational process might be changed, because I have no easy solutions. My central thesis suggests general lines of research that might lead to new methods.

In varying degrees, individuals need help in acquiring the tools of communication — how to read and listen to words, how to speak and write them. And everyone needs to learn how to resolve unconscious needs and conflicts so that they will not distort the work of the educated eye and ear and tongue and hand. But we do not need to be taught to think. In fact, thinking cannot be taught. The function of education is rather to show us how not to interfere with the thinking capacity that is inherent in the human mind.

The free creative velocity of our thinking apparatus is continually being braked and driven off course by the play of unconscious forces and, unhappily, by conventional educational practices as well. This is because we still try to educate largely by varying degrees of overemphasis on drill and grill, which interfere with the free play of preconscious processes both in the acquisition of new data and in the free utilization of that data.

When educators challenge us to tell them better ways, our only answer is that we can help them to undertake basic research on the educational process. This can lead to new practices that will leave our basic tools free for spontaneous use: basic research on how to impart new information without crippling the mind's creative potential.

The trouble with education: We still try to educate via drill and grill, wrongly assuming this to be learning. The point is that anybody can be bludgeoned into "knowing" that xenon is inert (which it is not), but that such bludgeoning cripples the mind's creative potential by discouraging true thinking.

This is a general educational problem; the fact that it has not been solved does not imply that we know nothing. We know, for instance, that thinking and learning are not performed consciously. Let us not minimize the importance of this fact, which challenges all traditional approaches to teaching — approaches based on the misleading assumption that we think and learn consciously.

The truth is that we learn preconsciously. The input of fragmentary perceptual data from the world around us is overwhelmingly preconscious. This constitutes an incessant subliminal bombardment, which goes on when we are awake and when we are asleep. The same is true for the most highly organized material. This has been demonstrated by experiments on the learning process under hypnosis, which indicate that our conscious intake is only a small fragment of our simultaneous preconscious intake, i.e., that process of intake and recall which occurs without conscious awareness. Furthermore, we know now that what the brain acquires preconsciously it also "processes" preconsciously, which is just another way of saying that all learning and thinking are preconscious rather than conscious processes.

This is not to say that our conscious processes are unimportant. They are important for sampling, checking, correcting, communicating, etc. But this is not thinking. The information we acquire consciously is never more than a weighted sample of the total input. This is one of the most basic facts of psychophysiology; and it has relevance for all educational processes. Yet it is a fact that is largely ignored by educators. So long as conscious sampling is mistaken for thinking, education will continue to neglect the great preconscious instrument of creative learning. Though we know that all effective recording, processing, and creating occurs preconsciously, we still must learn how to use this instrument. This is the major challenge that psychiatry brings to education.

We can learn new ways of educating only in schools that are both schools *and* laboratories. Just as research is an accepted function of every teaching hospital, research facilities must be attached to all leading public and private schools.

The best schools of tomorrow, those which provide the best education, will be those which carry on research every day and all the time in every detail of the educational process. I would like to see such research schools established first in some of our schools of science and engineering, because I believe that the unsolved problems of education can be studied with greater precision here than in other areas of education. Furthermore, if we can solve these problems in technical education, we will spark similar developments in every other kind of education.

Perhaps there should be a National Institute for Educational Research, paralleling the National Institutes of Health, with experimental schools and laboratories and funds for research. Toward that objective, for which there is an urgent need, the nation's leading schools of science and engineering now have an opportunity to provide a first step — by serving as laboratories in which this work can begin. Perhaps, at the same time, our technical schools can use this as an opportunity to do something about their high rates of attrition. Also it might help to solve a harder-to-measure problem, to which many engineering educators attest, namely, the fact that creativity so rarely survives the educational grind. If we succeed here, the technical schools will be the site of the discovery of new educational processes by which to protect the free use of our preconscious without crippling our creative potential.

I will conclude with an analogy. The freely gifted athlete can imitate with his body anything he sees anyone else do even once, and sometimes do it better than his experienced model. He will pick up a tennis racket, or skip rope, or put on skates, and in a moment will show natural precision and grace, better than the person whom he has been watching. He has a freedom to move and to put new movements together into new combinations. He does this with confidence, with bodily imagination, and without anxiety. He is like the naturally gifted artist who barely glances at something casually, and then with a piece of charcoal reproduces it automatically and faithfully. Then he can go further: He can take it apart into fragments and play with it, elaborating out of the initial stereotyped and literal image a new production out of the free play of his own creative fancy. There is much evidence that such effortless learning is the best, whether in graduate study or in our first days at school. We should search out methods by which to enable us to acquire our basic tools in this way. This is the challenge that psychiatry brings to technical education.

Lawrence S. Kubie

4 Freedom in Research

Frank M. Andrews and I are social psychologists. For the past several years, we have been studying eleven research organizations to find clues as to what makes for a stimulating research atmosphere: What kinds of relations among scientists, what kinds of personal motivations, raise or lower scientific achievement? In this essay and the one immediately following, where Andrews joins in the authorship, I want to tell you some of the things we have learned.

In this first essay, I will take up one of the major issues facing research organizations: the conflict between the needs of the scientist and the needs of his organization. Allison mentioned the issue in his earlier comments and I am sure you have encountered it yourself: Scientists want freedom, or *say* they want it. But the laboratory must accomplish its objectives.

We believe we can offer help — or at least some information — that may resolve this conflict between autonomy and coordination. The analysis I will discuss here, which was supported by the U.S. Army Research Office in Durham, was directed at answering this kind of question: If scientists say they want autonomy, does it follow that maximum autonomy results in maximum achievement, either scientific or practical? We believe this question must be answered before one can go on to resolve the conflict.

We felt that you cannot answer the question simply by asking scientists: "How much freedom do you prefer?" Instead, we attempted, first, to assess each man's level of autonomy;

then we examined his performance, mainly based on judgments by senior scientists who knew his work. And then we analyzed how well those scientists were performing, under different levels of autonomy or coordination.

Our work began with a questionnaire. Because we wanted to assess level of autonomy, we asked the participants to tell us who decided their scientific goals and objectives. A typical question — among 230 in all — is shown here:

QUESTION 29
Consider the choice of goals or objectives of the various technical activities for which you are responsible (either your own work, or work which you supervise or coordinate). Who has weight in deciding on these goals and objectives? Estimate the relative percent of weight exerted by each of the following, to nearest 5-10 percent.

	Percent of weight in deciding goals
Myself	______%
Subordinates	______%
Colleagues — other persons without supervisory authority over me	______%
My immediate chief	______%
Higher level technical supervisors in this organization	______%
Nontechnical executives	______%
Clients or sponsors	______%
Other:	______%
TOTAL (should add to 100%)	______%

If the participant assigned himself a large proportion of the weight in deciding his technical goals, we classified him as a man having high autonomy. On the other hand, if his chief or higher echelons exerted a large proportion of weight, we said that he did not have autonomy — that "coordination" existed.

Our next step was to measure performance. By "scientific performance" we mean any of several different measures of the scientific achievement of an individual. We did not attempt to measure performance of groups. We based our performance data on judgments by panels of senior scientists

— both supervisors and nonsupervisors — within each laboratory. We asked these senior people to rank all professional staff members with whose work they were familiar. This was the criterion: "Within the past five years, how much has each person's work contributed to general technical or scientific knowledge in the field?" These evaluations were made in confidence, to be sure. They were combined by a computer program into rank-ordering, from which a percentile score was established. In addition, we asked each participant to tell us how many technical papers he had published in professional journals over the past five years, how many patent applications, and how many unpublished technical manuscripts. You will see these data charted as we go along.

We have asked such questions of more than a thousand scientists and engineers in eleven research and development organizations. Five of these were industrial laboratories, including pharmaceuticals, glass technology, electronics, and electromechanical. Five were government laboratories, working in such areas as weapons guidance systems, utilization of agricultural products, and basic research in the physical sciences. The eleventh was a large university in the Midwest, where we studied seven departments, including biological sciences, math, physical sciences, and social sciences. These were not a representative sample of any universe of research laboratories, so we must be cautious about generalizing the results.

What did we find? Perhaps I should begin by telling you what surprised me most when we looked at some preliminary results. I had expected that people with high autonomy would have higher-than-average performance. But this was not necessarily the case. Further, one might expect that a scientist who allowed his colleagues to have some weight in deciding his goals would perform less well than the fellow who set his goals himself. In a very tentative way, it appeared that better performance occurred when influence on important decisions was shared with several persons at various levels.

But shared with *whom?* Might some combinations — self

+ colleagues, self + chief, for instance — work better than others?

One evaluation is shown in Chart One: for each participant, we went back to his questionnaire and combined the persons or groups affecting his decisions into four categories: himself, his colleagues and subordinates, his immediate chief, and executives (technical or nontechnical, including clients). Then we recorded how many of these "decision-making sources" exerted at least a slight weight — ten percent or more — on his decisions about technical goals.

These charts show that a technical man's performance is generally better when several "decision-making sources" are involved in setting his technical goals. A decision-making source may be the man himself, his immediate chief, his colleagues and subordinates, or higher executive levels. Men whose goals are influenced by one source — himself alone, for instance, or his chief alone — are grouped in "one decision-making source." Those whose goals are influenced by two groups are grouped in "two decision-making sources," and so on across each chart. Within these categories, the author shows how mean performance compares. Plotted vertically are several measures of performance, all taken over a five-year period. ("Scientific contribution" is based on judgments by senior scientists.) The trends we show are based on the author's data, but we superimposed the curves arbitrarily, so do not try to compare a point on "Papers" with another on "Reports." Rather, note what happens to performance as more sources are involved.

The first chart shows three measures of the performance of some 100 PhD scientists in development-oriented laboratories. (These were labs where scientists agreed that executives valued development of better products, in contrast to research-oriented labs, where executives stressed scientific publication.) When you compare the performance of these people — as judged by their senior scientists, and by the papers and reports they produced — you see that the more sources involved in affecting the scientist's goals, the higher was the average level of performance.

Let us see what happens to another group: nondoctoral people in development-oriented laboratories. Over half of these people, of more than 200 participants, were in engineering specialties. They worked in departments where fewer than one-quarter of the staff members held PhD's. Except for one deviation, as you can see in Chart Two, the same trend

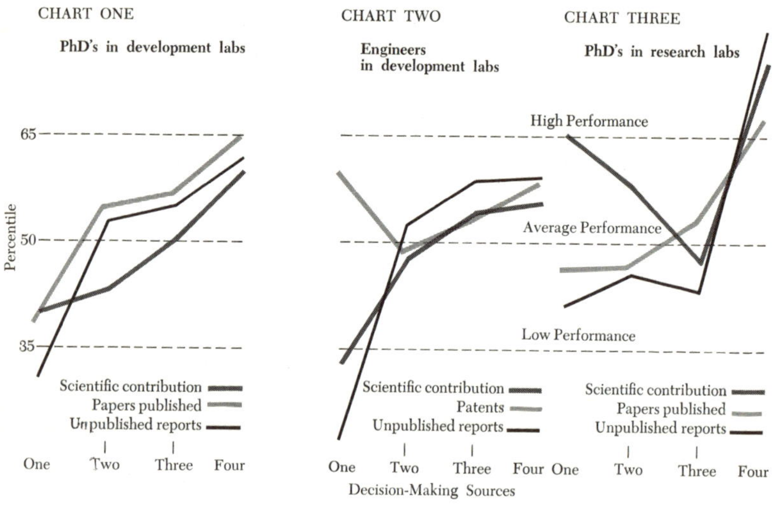

CHART ONE
For these 100 PhD's, in labs where executives place higher value on development of better products than on addition to scientific knowledge, it appears that highest performance is achieved when four decision-making sources are involved in setting the scientists' technical goals.

CHART TWO
These people also work in development-oriented labs, but where few members of staff are PhD's. Most of these 200 are engineers. Instead of papers published, as in Chart One, we plot "patents." Note that "patents" is highest among "one-source" engineers, i.e., those with autonomy.

CHART THREE
Of these PhD's, some 100 in all, about half work in government, the others in university departments. Publication of knowledge is more highly valued here than the creation of better products. Why do "one-source" people contribute more? Perhaps those autonomous PhD's talk more than they write.

existed as with the PhD scientists: the more sources involved, the higher the performance.

Up to now, we have talked about development laboratories. Did these same trends hold for laboratories that were research-oriented (that is, labs where executives stressed scientific publication)? In Chart Three, we see some similarities to the earlier evaluations, plus one puzzling difference. In terms of producing papers and reports, these research PhD's performed in about the same way as the development scientists and engineers: people whose goals were affected by four levels published more papers and wrote more reports than those who were influenced by fewer sources. But the puzzling difference was the plot of scientific contribution: the high performers were in the four-source group and the one-source group. I can only speculate on this. Perhaps the high-performance, single-level scientists talked more than they wrote; perhaps these high performers were discussing their work with colleagues, rather than taking time to write it up.

But let us come back to the main result and examine some objections. You might argue, for example, that scientists whose

Relations with colleagues and supervisors related to productivity. Mutual collaboration was best, complete direction worst.

work is scrutinized by several levels are likely to be senior people; they are more productive, *not* because they are involved with more people, but because they have greater experience.

Another argument one might raise is that better people become supervisors and supervisors' work is scrutinized by several echelons. We did some further analyses that tended to rule out these explanations.

One might still argue that the fruitful worker is the one who gets the attention from other sources — a simple and gratifying explanation; many scientists like it. But I shall argue that the involvement of one's colleagues, research executives, or client representatives can actually stimulate high performance.

I said earlier that I had expected high autonomy to go with high performance. And this belief has carried over into my teaching. For several summers, I have taught an introductory course at Michigan in survey research methods. Because I believe in autonomy as essential to education, I had always asked each student to design a hypothetical survey on a topic of his choice. But recently I tried something different. I asked the class to work on a single project, one that was of vital concern to our community: Why did the voters of Ann Arbor first reject, and then accept, requests from the Board of Education for higher taxes to support a larger school system?

Members of an Ann Arbor citizens committee acted as our client. They presented the problem to the class and raised a small sum to defray our costs. Within the class, committees were formed to decide what they needed to find out, construct a questionnaire, plan the analysis, and so on. These class committees interacted vigorously and reported often. I studied their drafts as we went along and suggested changes in light of my professional experience. Such changes were made (I hope) by mutual consent with each committee chairman. The citizens committee came by to hear the reports of the class committees as the survey progressed.

What was the effect? The quality of the class committees'

work was first-rate. A colleague of mine, who knew one of the students, told me their out-of-class involvement was "fantastic."

As I thought about this, it struck me that these class procedures were exactly in line with our previous findings. Four levels were involved: the individual student, with considerable autonomy to initiate and follow up his own ideas; his colleagues, who interacted vigorously on committees and in the class; myself, as the immediate supervisor; and the citizens committee, as the eager client. I am convinced that the multisource approach provided far more excitement and challenge than had been the case with my summer classes in previous years.

Let us come back to the original dilemma: Is the scientist's freedom eroded as more people influence his decision-making, his goals?

This brings up the question of *influence* or control. To what extent can a scientist influence key decision-makers? My hunch is that if the scientist retains substantial control over powerful decision-makers, then the involvement of these powerful sources in his decisions will *stimulate* the scientist.

QUESTION 31

Please write the name of the one person or group (other than yourself) who has most weight in choice of your work goals. . . . To what extent do you feel you can influence this person or group in his recommendations or decisions concerning your technical goals?

CHECK ONE

____ I can exert almost no influence
____ A little influence
____ Moderate
____ Considerable
____ Great influence
____ Irrelevant, since no one else affects my choice of goals

I base this on the following findings that emerged from our study. We wanted to measure influence, so one of our questions was asked in two parts: (1) Who has most weight in choice of your work goals? and (2) How much influence do you have over this person or group? With answers to this question, we could divide our participants into two groups: those with high feeling of influence over important decision-makers and those who felt lesser influence. Then, within these groups, we examined the relationship between number of sources and performance. Our findings are shown in this next trio of charts.

In Chart Four, we see performance plotted for those with high influence and those with lesser influence, among PhD

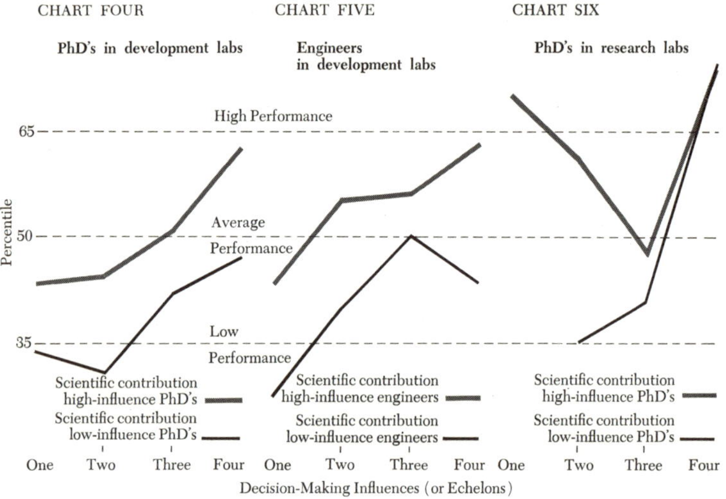

CHART FOUR
The PhD's in Chart One rated themselves in terms of personal influence (see Question 31). Here we see that maximum performance occurred when these PhD's had both high influence and the involvement of several others.

CHART FIVE
Here are "high-influence" and "low-influence" engineers. We see again that those with high influence show a rising trend in performance with more decision-making sources. Those with low influence seem to perform best with three sources.

CHART SIX
What this chart may imply is that the low-status scientist in research — one with low influence — should not work in isolation, but that the scientist who commands respect — high influence — works most effectively when he is on his own.

scientists in development laboratories. Perhaps not surprisingly, we found that the high-influence people rated higher on performance. But note that with both groups the same upward slope appeared as in those earlier charts: the more involved, the better. This was particularly true for high-influence scientists. Retaining a substantial voice in their goals, these people appeared to benefit if their planning was shared with several other groups. Maximum performance occurred when the scientist had *both* high influence and the involvement of others. On the other hand, if he lacked influence, multilevel involvement was helpful, but less so.

In the fifth chart, we see data for engineers in development laboratories. Again, those with high influence showed a rising trend in performance as more sources were involved. For engineers with less influence, three levels rather than four seemed desirable. Why is this so? It may be, as in the case of the nondoctoral people back in Chart Two, that the low-influence engineer is less secure. To discuss his work with a wide audience may threaten as well as stimulate. Thus, three sources — himself, his chief, and a third party, such as a respected colleague — seem optimal.

In the sixth chart, we have data for PhD's in research laboratories. The high-influence scientists showed the same trend as in Chart Three. It may be that those men who command respect, i.e., high influence, can work effectively with only one or two levels. On the other hand, it seems that low-influence scientists in research should not work in isolation, but should contact at least two other sources besides themselves.

These results remind me of some observations I made in 1955, in a study of an industrial petrochemical laboratory. It seemed to me the company had been successful in building strong motivation. It had done this by giving a large measure of individual responsibility to its technical people, one of whom used an appropriate term to describe this process: he called it "controlled freedom." He described its working as follows. A general problem area was sketched out for him; he was shown what mountain to climb, and then it was up to him to get to the top. But he was not ignored: He was encouraged to sketch out a program and to make regular progress reports. Further, he was encouraged to maintain close direct contact with others who had an interest in his work and to whom his work was important. Sometimes, this meant conferences with directors in his division or in a higher division. Sometimes it meant contacts with customers or manufacturers. These meetings served to keep the research man on his toes, to funnel information to him that often was useful, and particularly to give his upper-level executives a chance to appreciate what he was doing. The feeling that others are interested in your work is an excellent way of sustaining interest in it yourself.

Thus far, I have simplified the data on Question 29 — shown earlier — by speaking only of the *number* of persons or groups having some weight in determining a man's goals. I have not told you who was represented in those various groups. When the chart said "one decision-making source," for

CHART SEVEN

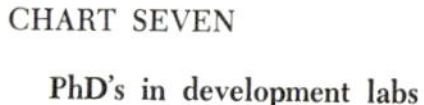

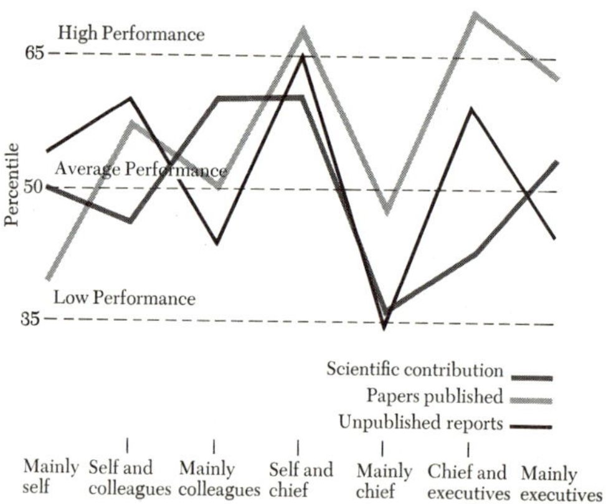

In these next three charts, we see how various influences affect performance. Among PhD's in development labs, it appears that performance is best when the scientist's technical goals are decided by his chief and himself; performance is lowest when the chief alone decides scientist's technical goals.

instance, did it mean the man himself? his boss? a top executive?

To answer such questions, we used another scoring system, which you will see in Chart Seven. I know it looks complex, but please bear with me as I explain its meaning. If the participant said on Question 29 that he alone was the only source, with at least 30 percent weight in setting his goals, we classed him in the "mainly self" category on Chart Seven. If he said that he and his colleagues both had at least 30 percent weight, we classed him in "self and colleagues." And so on.

In other words, the "mainly self" category was essentially a condition of autonomy. Within this category we can see how

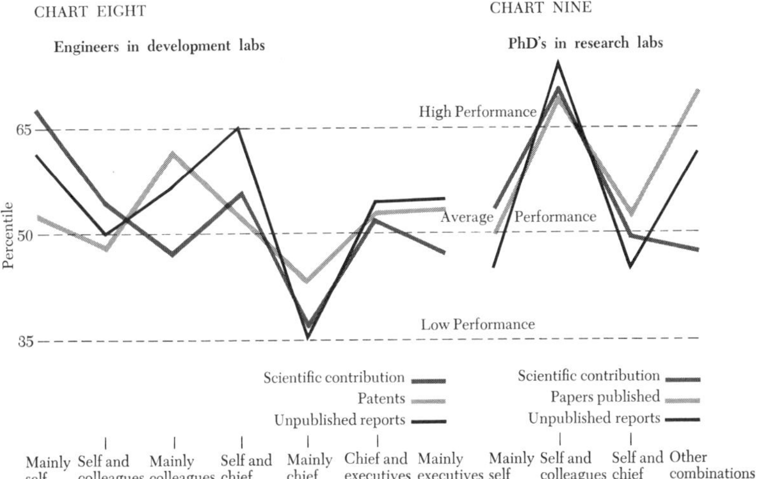

CHART EIGHT
Engineers are different. Unlike the PhD, the autonomous engineer (who sets his goals himself) is a good performer. But engineers resemble development-lab PhD's in two categories: performance is high when "self and chief" set goals, but it sinks when "mainly chief" decides what the goals shall be.

CHART NINE
The research scientist usually sets his goals himself, or jointly with colleagues or chief. Hence the brevity of this chart. Unlike the development scientist (Chart Seven), the research PhD seems to perform best when "self and colleagues" establish his technical goals.

performance compared with performance in the other categories. How *did* it compare? The chart indicates "mainly self" was not very fruitful, by any of the three performance measures. The most favorable combination was the condition of "self and chief," where the scientist and his immediate supervisor had a large mutual voice in setting his goals. And just as clearly, the condition of "mainly chief" was unfavorable by all three performance criteria.

For other combinations, the picture was mixed. When "self and colleagues" set the goals, many papers and reports were written, but to judge from the professional evaluations of senior scientists, these were of mediocre scientific value. This is an example of a "leaderless team." On the other hand, when "chief and executives" set the goals without the scientists' participation, a similar pattern emerged: many papers and reports, but of mediocre value. Here is an example of a team without members. But let us beware of reading too much meaning here, for the discrepancies may result from a variety of unexamined factors.

We made the same analysis of the engineers in development laboratories, as shown in Chart Eight, and found some similarities with the scientists, along with some differences. For example, like the scientists, the engineers' performance was good when "self and chief" set goals; and performance was not good when the goals were set by "mainly chief." But in contrast to the scientists, the autonomous engineers ("mainly self") were highly creative.

Corresponding data for PhD's in research laboratories are shown in the last chart, Chart Nine. Only three of the ten possible "major weight" combinations occurred here with any frequency, and hence this chart has fewer categories. We might expect this: Decisions were made largely by the research scientist himself, or jointly with his colleagues or chief. The most fruitful pattern was such that "self and colleagues" set the goals. (In contrast to PhD's in development, goal-setting by "self and chief" was not productive, by whatever criterion it was measured.)

What puzzled us was the fact that autonomy, enjoyed

by half the sample, yielded only average output. How do we reconcile this with the data we showed you in Chart Three — where one or two decision-making sources, or levels, seemed to be better than three? The explanation may lie in how many other sources were *slightly* involved: When we looked at the scientists who *mainly* set their own goals — 30 percent weight or more — we found that only one in three had sole control (no other group as much as 10 percent). These were the high contributors in the third chart. The other autonomous scientists shared their decisions slightly with one or two other sources. I am tempted to make this tentative conclusion: It is the rare scientist, even in a research laboratory, who can be creative mainly from his own resources.

If I have provoked you with my comments, no doubt you are eager to challenge me with questions. I will conclude, then, with the sorts of questions I hear nowadays from management people who have heard these results — but are skeptical. If you like to jump from chapter to chapter when you read a book of essays, you may already have read Rensis Likert's piece, which appears in the next section — and since his observations relate to mine, you may want me to answer the following:

Your boss, Rensis Likert, says that traditional systems of management do not work with creative people in science. How do your ideas jibe with his?

Likert's ideas in his supervision piece are based partly on results from nonscientific organizations. And — to my own surprise — several of these ideas fit remarkably with our results. In his *New Patterns of Management,* Likert describes a national service organization in which employees rated the amount of influence they felt was exerted by management, and the amount exerted by employees themselves. In high-producing departments, Likert found employees exerted more influence, but also that other sources exerted influence too. In short, there was more *total* influence than was the case with low-producing employees. The old view had been that *quantity* of influence is fixed: If subordinates had more, superiors would have less. What Likert showed was that high-producing managers had actually increased the size of the "influence pie" by means of the leadership processes they use. And our colleague, Arnold S. Tannenbaum, shows similar results in labor unions, Leagues of Women Voters, and business firms.

You say that exposure to several sources is stimulating. Isn't it possible that as the scientist matures he makes contact with more levels and also produces more? Couldn't this explain your results?

We tried to rule out this explanation by adjusting our performance data to hold education and experience constant. In the data you have seen, we have already factored out systematic effects associated with possession of the PhD, with years of experience, and with length of time in the organization.

What I really mean is not just experience, but overall ability. Cream rises to the top of the bottle. Better people become supervisors — and supervisors have to deal with several groups. Can you disprove this explanation?

We broke our sample into two levels, supervisory and nonsupervisory, and repeated our analysis within each level. For the most part, the same trends reappeared. The possession of supervisory status did not account for our results.

Well, just for the sake of argument, suppose we grant that exposure to several sources really is stimulating. Isn't this because the scientist is getting information, and not because these other people are telling him what to do?

Maybe so. Information, after all, *is* influence. And it can flow in both directions. If the scientist can tell higher echelons what he has found out, he can influence their decisions. If they present their needs and concerns, they let him make more rational choices. They influence him by information rather than by fiat.

Your "levels" bother me. Research men hate being layered. Are you advocating a tall hierarchy with many levels of review?

Definitely not. When there are several levels of review, each with an ax, the scientist has no influence. What we advocate is a chance for *face-to-face interaction* between the scientist and other significant people in his research and development system.

Aren't there too damned many meetings already?

Watch your language. Small committees, properly used, can maintain an atmosphere of excitement and challenge. Perhaps you could limit the demands on time by giving the scientist an option: Let him report verbally a couple of times a year to a small committee, instead of writing a progress report.

Sounds like one more administrative chore, serving on that committee. Whom would you recruit?

I am inclined to conclude that it stimulates a scientist to spend some time, but not a lot, doing administration or teaching that is related to his work. Why not give every scientist in the lab, junior as well as senior, a chance to serve at least once a year on at least one review committee? Andrews has found that scientists and engineers who spent full time at their research and development tasks were less productive than those who spent, say, three-quarters or half time.

You talk about the low productivity of scientists whose goals are set by their immediate chief. It is obvious to me what is happening: If the chief has a subordinate who isn't very good, he has to make most of the major decisions himself.

Maybe so, but if you want to stunt him permanently, be sure to go on doing just that. My feeling is that if this scientist can see two points of view, if he gets input from colleagues, or other supervisors, or from clients, this can stimulate his growth. But if his directions come from one man only, his boss, who also evaluates his performance and determines his pay, this can be deadly. I cannot think of a better way to stifle independence of mind.

Here is an instance in which the director of the laboratory can be imaginative. Suppose some of his section chiefs—

brilliant men — allow little leeway to their subordinates. What does the director do? I doubt that supervisory training will help. But perhaps the director should see that the section members get the opportunity to interact with other groups — with colleagues, higher supervisors, outside sponsors. What I have in mind are the kinds of conferences I described earlier, in the petrochemical laboratory. In such interaction is potential for growth.

Donald C. Pelz

Diversity in Research 5

Another area we have been exploring in our investigation of questions relating to the research and development atmosphere is this matter of diversity: Is it a good idea for the technical man to be involved in a variety of activities? Or is he better off to concentrate on one thing and try to achieve real proficiency there? In the previous essay, you heard about some of our work dealing with the question of freedom; now we want to talk about the content of the technical man's work:

What is the most fruitful way in which he can spend his time?

Should the research man concentrate exclusively on research . . . and the development man on development?

Should he strive for "depth" in a few problem areas, or for "breadth" in several areas?

These are the kinds of questions we took up in this study, and within these question areas we tried to identify all factors that were influencing our results. For instance, is the "depth" approach better in certain kinds of laboratories? Is "breadth" — or "depth" — to be preferred at certain stages in the individual's career? And we looked at related questions: How much of his working day should be spent strictly on technical tasks, and how much, if any, on administration, teaching, communication? How many projects can he work on profitably at one time?

On the next pages, you will see the results of our exploration and our interpretation of those results. As you scan our charts, keep in mind that we did not simply *ask* people what they preferred — in terms of "breadth" or "depth," for example. Rather, we asked how they were actually spending their time. Then we examined each man's performance.

How does one measure performance? We relied mainly on judgments by panels of senior scientists within each laboratory, both supervisors and nonsupervisors. These senior scientists ranked all professional staff members, with whose work they were familiar, on the criterion: "Within the past five years, how much has this man's work contributed to general technical or scientific knowledge in the field?" In the charts on the following pages, you will see these judgments plotted as "scientific contribution." Rankings from various judges were combined into a single score for each of the men included in our study. We also scored each man on the number of papers he had published, his patents or patent applications, and his unpublished reports or manuscripts — all within that five-year period.

Now for some findings. One of the first things we looked at was — what shall we call it? — distractions on the job. We asked each man to tell us what percent of his total work time he spent at each of three things:

Technical work — whether his own work or technical supervision and collaboration.

Administration — including communication with higher-ups or outside groups.

Teaching or training.

With these data in hand, we then compared the performance of our participants who spent various proportions of their work time in these activities.

What surprised us?

We should be frank here and state a couple of our preconceived notions. For instance, we felt that some degree of teaching probably enhanced scientific achievement. And we suspected that time in administration probably hindered such achievement. But just how much administrative distraction could the scientist tolerate, we wondered, before his productivity suffered?

Now let's talk about the first surprise.

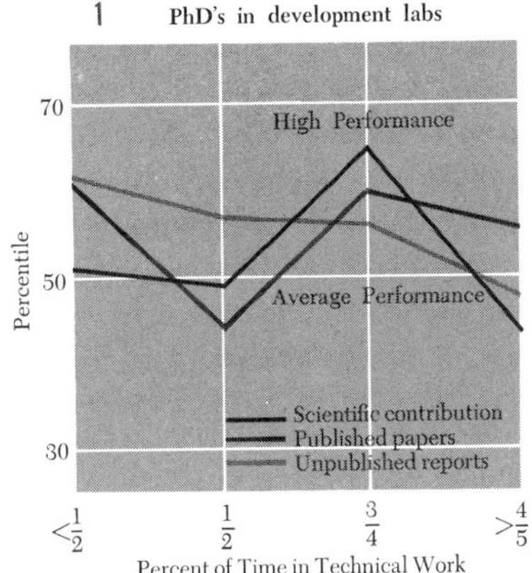

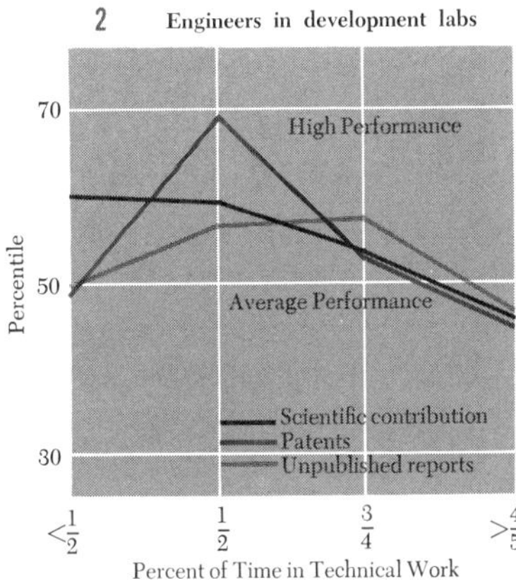

Certainly the results of this first chart, and the two charts following, were not what we had expected. In this chart, we show data for about 100 PhD's in development-oriented laboratories — labs where scientists agreed that executives valued product development, in contrast to research-oriented labs, where the emphasis was on scientific publication. It appears here that scientists who spent essentially full time in technical work — and hence little if any time in administration — were less effective as scientists than those spending about three-quarters time in technical work.

And we see a similar trend in Chart 2, among some 200 nondoctoral people in development-oriented laboratories. Over half of these people were trained in engineering. They worked in departments, both governmental and industrial, where fewer than one-quarter of their staff members held PhDs. Optimal performance here occurred among people who spent only half time in strictly technical activities. Instead of "published papers," we plot "patents." Mean five-year patent output at 70th percentile was roughly 3.5 per man.

Plotted vertically on these three charts are several measures of performance, all taken over a five-year period. Scientific contribution is based on judgments of senior scientists. Means outputs of published papers and unpublished reports are based on a logarithmic transformation. All scores are adjusted to remove effects due to length of work experience since degree, and have been superimposed so that 30th and 70th percentiles for all three scores coincide. In first chart, for instance, a group whose mean score on papers fell at 30th percentile would have produced roughly two papers per man in five years, while a group whose mean fell at 70th percentile would have produced about 16 papers per man.

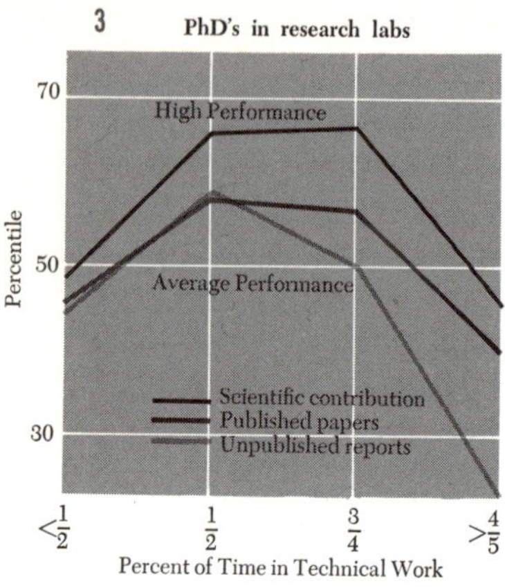

Among 75 PhD scientists in research-oriented labs, we see in Chart 3 that those who spent only half or three-quarters time on strictly research activities were more productive scientifically than those who spent full time. Our university departments are included here — also half the departments in government laboratories.

You may be coming to the suspicion now — as we did — that diversity was critical. By several criteria, which we shall examine, we shall see that technical people were most effective if they had several interests, less effective if they were highly specialized people.

Now you are wondering: Were the people who spent full time in technical work simply junior members who had not yet produced much? Were those who spent three-quarters or half time in technical work the senior people, or the supervisors — people who not only produced more but who also had administrative chores thrust upon them? We wondered

"The scientists we have examined were most effective if they had several interests, less effective if they were highly specialized."

too, so we checked this point carefully . . . and we found it did not explain the results. We found that among nonsupervisory researchers, those spending full time in research were *simply less effective*. This was true of junior people *and* senior people.

This might have puzzled us more than it did, but we recalled that similar results were found several years ago in a national survey of physiologists. A colleague of ours at the Survey Research Center, Leo Meltzer, discovered that full-time researchers in physiology published less than those spending three-quarters time, regardless of academic rank.

What about teaching versus administration?

Among these research-lab PhD's, the time spent in "nonresearch" was given either to teaching or to administration. Sometimes it was given to both. On the other hand, the people in the two earlier charts — the development-lab people — spent most of their "nontechnical-work" time in administrative tasks. We shall say a word later about the distinction between these — the "good" effects of teaching and the "bad" effects of administration. Another surprise.

We had suspected that spare-time teaching was probably more stimulating than spare-time administering, so we went on to check this hunch:

In development labs, we could compare only those people who did *some* teaching with those who did none at all, because little formal teaching was done in those labs. And we found, among engineers in development, that some time in teaching proved a slight advantage. However, among PhD's in development, it did not seem to matter whether a man taught or not: Teaching did not affect his performance.

In research labs, on the other hand, a most curious result appeared. Among nonsupervisors, some taught during their nonresearch time; others did some administering. Those who spent their off-time mainly in administration were *more* productive in their research than those who spent their nonresearch time mainly in teaching.

Is it possible that some administrative activity actually stimulates scientific performance? Possibly "administration" in mild doses is not a mere exercise in pushing papers. Rather, it exposes the man to fruitful ideas outside his immediate problem area.

The next three charts show another piece of analysis: Is there a connection between the *number* of specialties a man is engaged in and his performance?

We asked our participants the following question:

"Within a discipline or field, an individual may develop an area of specialization—a content area about which he knows a great deal. If you have such areas of specialization, please list them below. . . . (Limit to areas in which you are currently active.)"

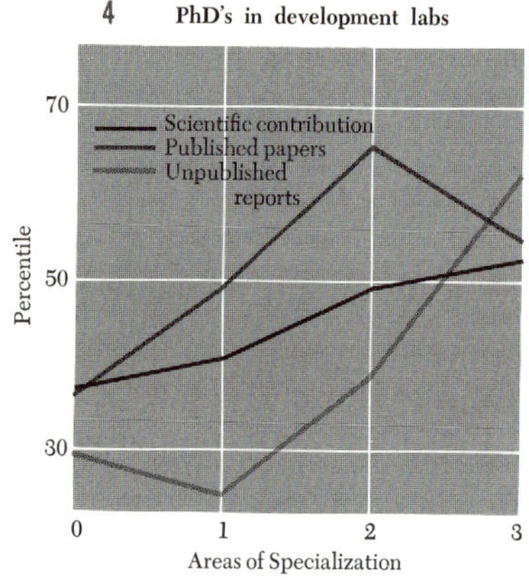

Now look at Chart 4: Here once again are our 100 PhD scientists in development-oriented labs. The *more* areas of specialization reported by these people, the higher was their scientific performance. The only exception here related to published papers: More papers were published by the two-specialization people than by those having three. We don't know why.

Our engineers in development labs (Chart 5) performed similarly: Those with two or three areas of specialization performed better than those with none or one.

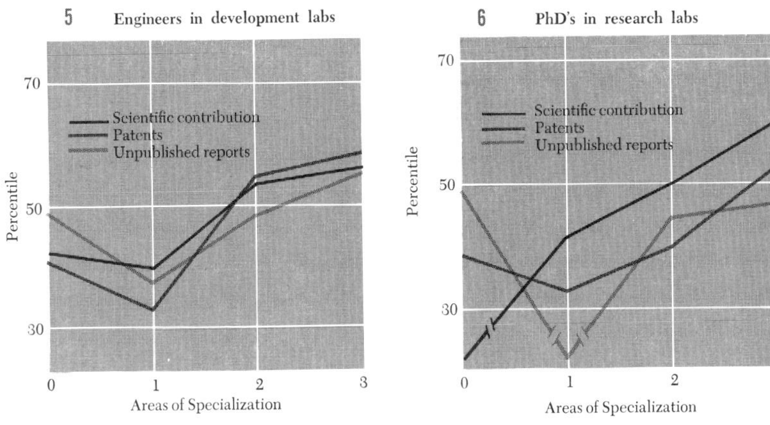

And in Chart 6 we have a similar trend for the research-lab PhD's: Except for those people who told us they had "no" areas of specialization — and there were very few who claimed no specialty — we see that performance went up as areas of specialization went up.

Before we go on to further analysis, let us try to anticipate your questions: Might it be, you are asking, that as the scientist matures he acquires more specialized areas, and also produces more? Hence, might the relationship between perform-

"Lack of breadth appeared to be a serious handicap to juniors, less a handicap to seniors, and least inhibiting for supervisors."

ance and specialization be due simply to length of experience? Later, we will show what happened when these data were re-examined by career stages. Here, let us answer that such analysis did not wipe out the results. If anything, the analysis did precisely the reverse of that.

Thus far, the picture suggests that diversity is an aid to the scientist or engineer in his research or development tasks. But let us check this with some other evidence.

QUESTION 14

The technical work of scientists and engineers covers a broad range of activities. At present, about what percent of your time (other than teaching) is directed toward each of the following purposes (either your own work, or work for which you are responsible)? Enter nearest 5-10%. FILL ALL SPACES.

	Percent of work
I. Research (discovery of new knowledge either basic or applied)	____%
A. General knowledge relevant to a broad class of problems	____%
B. Specific knowledge for solution of particular problems	____%
Subtotal should equal item I . . .	(____%)
II. Development and invention (design of particular products or processes; translating knowledge into useful form)	____%
C. Improvement of existing products or processes	____%
D. Invention of new products or processes	____%
Subtotal should equal item II . .	(____%)
III. Technical services to help other people or groups . . . . [Including testing, analysis by standardized techniques, consultation, trouble-shooting]	____%)
IV. Other purposes (specify):	____%
Total time should add to 100% . .	(____%)

The question shown here (Question 14) enabled us to explore another line of analysis: How should the technical man allocate his time among various research and development functions? When our participants had answered this question, we knew how many *different* functions each man was engaged in. We took the five categories shown — "general" (or basic) research, "specific" (or applied) research, product improvement, invention, and technical services — and we recorded the number of these on which the man spent 6 percent or more of his time. The next three charts show the results.

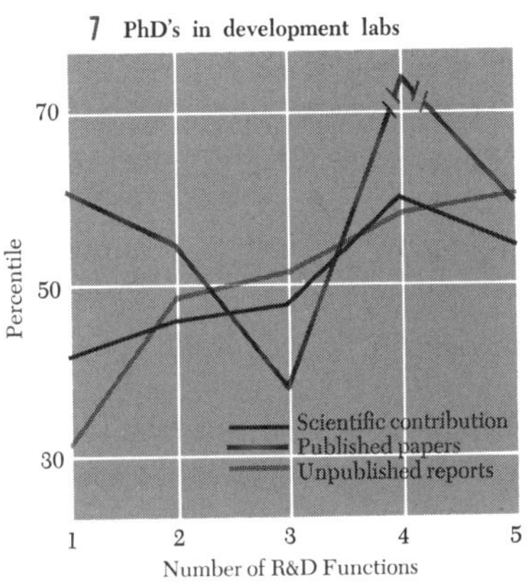

For PhD's in development, the general tendency was clear: The more kinds of research and development functions the scientist was engaged in, the better was his performance. True, the man who was engaged in only one function published more papers than his colleague who was engaged in two or

three — but note that his scientific contribution and his report output were lowest.

For engineers, maximum performance occurred among men who were engaged in five different activities. It was minimal among engineers engaged in only one.

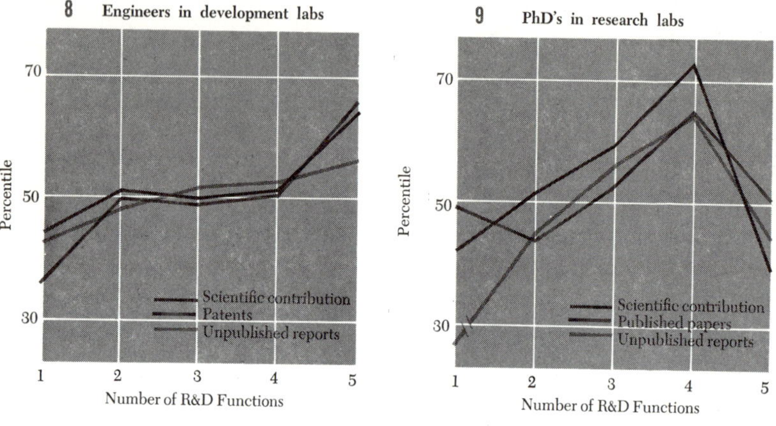

Among PhD's in research — as with development PhD's — performance increased as functions increased, up to a maximum at four functions. It is interesting to note the drop from four functions to five: In research there is profit in diversification, but also danger; carry it too far and you become superficial.

We believe the generalization to be drawn from this trio of charts, as from those shown earlier, is that the more productive people are those with moderate diversity in the content of their work.

The next chart compares the performance of development-lab PhD's who spent different proportions of time in research. (Time spent is taken from "I. Research" in the questionnaire we quoted a moment ago.) A few of these PhD's who told us they spent *no* time in research were highly productive individuals. Among the others, however, the optimal proportion of time in research was about one-half to three-quarters. Those who devoted very little time — shown at 6 to 20 percent in the chart — or a great deal of time (more than 80 percent) were less effective, as measured by "scientific contribution" and unpublished reports.

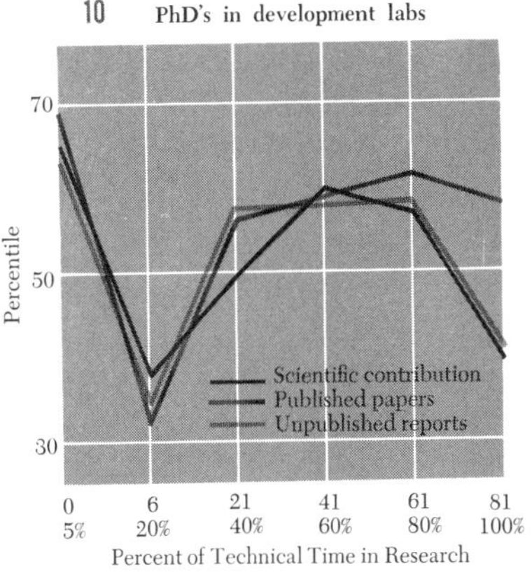

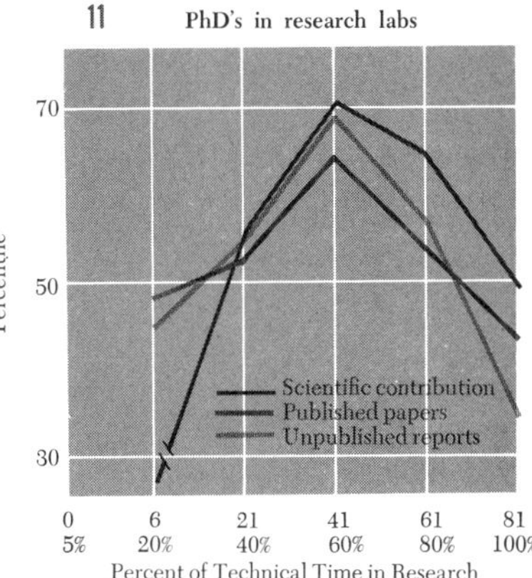

PhD's in research labs followed a similar trend. The best performers appeared to be those who spent about one-half to three-quarters time in research. Those who spent nearly full time — or very little time — in research were definitely less effective.

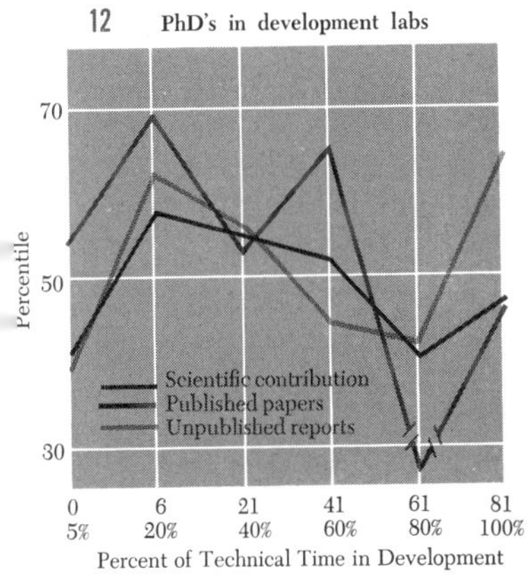

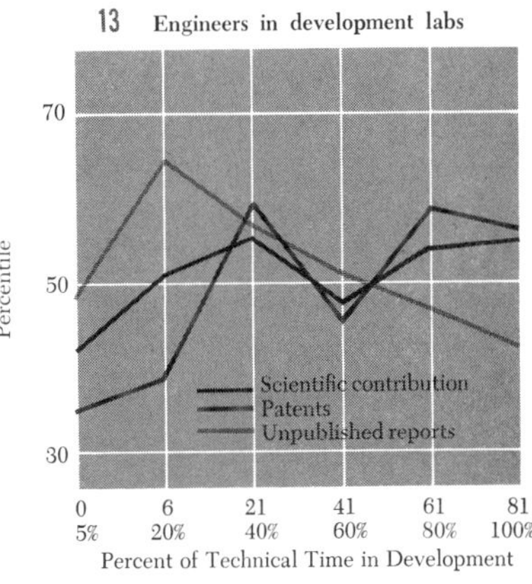

Now let us see how performance varies with time spent in development. (Time spent is taken from "II. Development and invention" in that same question we showed earlier.)

For development-lab PhD's, we see that a few individuals performed fairly well if they spent nearly all their time in development. (These were probably the same people who spent *no* time in research — in Chart 10.) But for most of these development-lab PhD's, the best performance occurred among those who spent from one-eighth to one-half time in development activities. Even among PhD's in research, for whom we don't show data on this point, performance was somewhat better among those who spent an eighth to a third time on development — rather than *no* time or considerable time. (Very few of these research PhD's spent more than half time in development activities, as one would expect.)

The development-lab engineers showed a modification of the same trend: Some of these men tended to be productive if they spent about one-third time in development, but others contributed or patented more if they spent three-quarters to full time.

The third category we included in that earlier questionnaire was "III. Technical Services." We are not going to bother charting it, but two points regarding it are worthy of brief comment: Few PhD's spent more than a third time in technical services, but those who gave *some* time were slightly more productive — both in research and development — than those who gave none. And the same was true, generally, for engineers.

So we see again that in both research and development, the most effective scientists and engineers — with some exceptions — did not concentrate exclusively in research, development, or technical services. But neither did these most effective people totally avoid these categories. In all three areas, some time was better than none at all.

If it is true that men who diversify their activities are better performers, does this hold true at all career levels? Are there stages in a man's career when he should dig deep, other stages when he should broaden out?

To answer such questions, we first had to define a set of career levels that would have analogous meaning across different kinds of laboratories. The labs in which our study was made did not have comparable job ladders or grades, of course — some were industry, others government or university labs. By looking at job titles and salaries assigned to these grades, plus education, length of service, and number of subordinates, we were able to establish rough parallels among them. Then we could go ahead with our comparisons. You will see these in the next three charts.

Supervisors included members of industrial or government labs who headed a section or larger unit, and university department chairmen and full professors.

Seniors included mature nonsupervisory investigators with substantial responsibility.

Juniors included younger scientists or engineers. The typical man was in his early thirties, eight years beyond the BS.

The apprentices were nondoctorals at the bottom of the professional ladder, mainly recent college graduates in routine tasks.

14 All PhD's in research and development labs

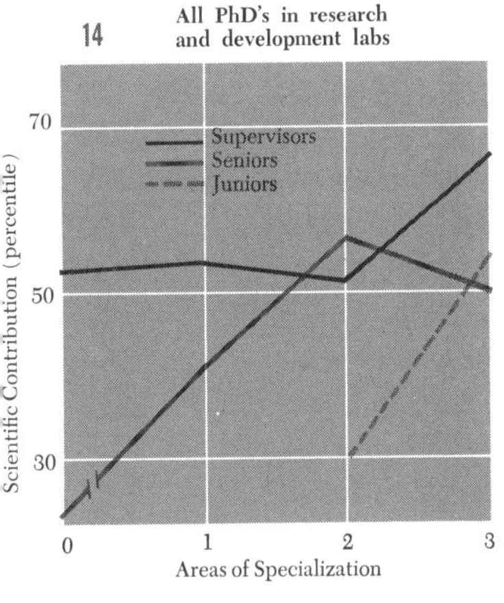

15 Engineers in development labs

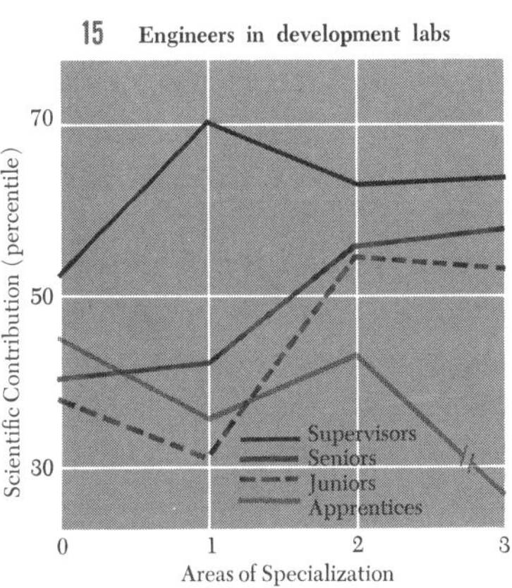

Among PhD's having at least three areas of specialization, there was not a sharp difference among those at each career level. But as we move to the left on the chart, we see that difference among levels increased: Supervisors did not seem to be especially disadvantaged, but possession of fewer specialties was a handicap to younger people. Lack of breadth appeared to be a serious handicap to juniors, less a handicap to seniors, and least inhibiting for supervisors.

For engineers, it appears that breadth of specialties was especially important among junior and senior men. For apprentice engineers, however, specialization may actually have helped. And having a single specialty may be beneficial for supervisors.

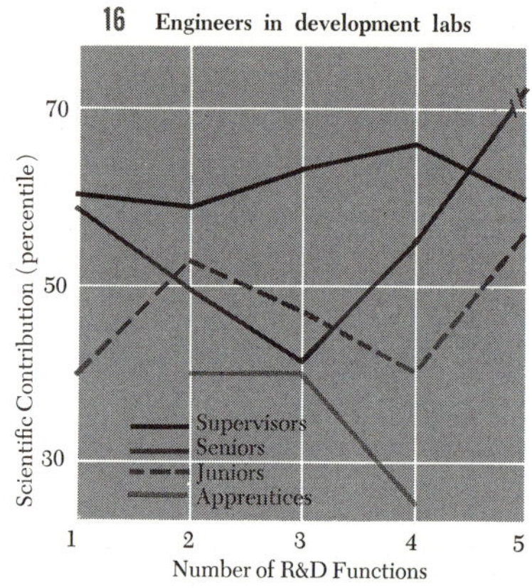

Another interesting picture appeared when we compared the scientific contribution among various career levels in relation to the number of research and development functions performed. For instance, we see that apprentices, as well as some junior engineers, performed better when they limited their attention to two or three R&D functions. But some juniors did well when covering five. And among senior engineers, this bimodal effect was heightened: The best records were made either by those senior engineers who concentrated on a single function or by those who diversified over four or five. We can almost see these men taking one of two routes to success: either the route of the narrow specialist or that of the broad generalist.

At the beginning, we raised a question concerning the number of projects the technical man could work on profitably at one time. Our charts did not answer this question, so let us say what we found — and then we will go on to our conclusions: For most groups, we found a very slight tendency for scientists and engineers to perform better if they were working on two or three projects at a time. (The research PhD's were an exception here: For them, it seemed to make no difference whether they were engaged in one project or more.) But for all groups, we conclude that the man performs best when he is utilizing two or three different *skills,* and when he is faced with *both* basic and applied problems in his work. It does not seem to matter whether he thinks of his work as organized around one project or several, so long as there is this *mix* in his activity.

One other generality ran through all groups: The man who worked the standard eight-hour day performed less well than the man who worked nine hours or ten. But it did not follow that the longer the hours the better the performance: the nine- or ten-hour-day man performed better, generally, than the fellow who worked eleven hours or more. Once again, there was a hint that *excessive* dedication was not healthy and that all work and no diversity was making Jack a dull scientist.

What are the implications here for the administration of research organizations?

One conclusion we come to is this: Complete protection of

"We can almost see these engineers taking one of two routes to success: either the route of the broad generalist or that of the narrow specialist."

technical staff from administrative duties may be a mistake. Our information suggests that technical people actually *benefit* from a mild exposure to administrative tasks. And, indeed, this information jibes with the findings we described in the earlier chapter on research freedom: The technical man may benefit if his colleagues are involved in setting his technical goals. The information we have presented here suggests that the technical man might also benefit from such administrative tasks as participating in conferences and serving on review committees — affairs at which he and his colleagues periodically discuss the future directions of the laboratory, and at which each man may be asked periodically to outline the directions of his work. We believe the membership of such review committees should not be limited to a few supervisors. If this responsibility were spread around among all members of the professional staff, it would add zest and breadth to the research climate.

Our second conclusion is that younger scientists and engineers should be stimulated to develop several specialized areas. And finally, we believe the research director should *not* assign certain groups exclusively to research, others exclusively to development. Rather, he should encourage groups and individuals to tackle both "pure" and "applied" problems.

Now if you are skeptical regarding our results, you are probably thinking: "I know they've not claimed outright that diversity stimulates — they've only *hinted* at it. But I say that effective scientists and engineers are just naturally curious people. The bright, productive ones are noticed and consulted about all sorts of problems. And *that's* why they develop more specialties and get into different kinds of research and development activities."

Maybe you're right. With data llike ours you can never prove one causal hypothesis over another. But why do you *insist* that it is *always* the individual and not the environment that makes the difference? On this premise, the man is either inherently capable or he is not, and the outstanding ones are those who are drawn into environments we find linked with achievement.

But maybe something else is happening. Maybe the best people are deliberately exposing themselves to the kinds of contacts that will stimulate them. If this is so, is it not possible for the good — but not outstanding — scientist or engineer to learn a trick or two? If he were in the kinds of situations that typify the best scientists — or if his manager saw that this happened — might not this "next-best" man actually boost his own achievement?

"What do you suggest I do?" you may ask. "I'm a research manager: Do I assign each of my people to three more projects? You don't expect me to believe I can spoon-feed diversity?"

No — that tactic would be artificial. Note that it does not matter how many *projects* the man works on. The important thing is the number of *skills* or *specialties* he brings to bear . . . and you don't develop these skills or specialties overnight.

But what you, as manager, *can* do is this: The next time you need to probe a specialized area — whether a one-week review or a one-year pilot project — give the job to a man (or a small group) who is working in a related area. Try to interest him in exploring this one. *Don't* give the job to the man who already is a specialist in that area. The man in a related specialty will dig into the field with new zest and excitement. He may develop fresh ideas that experts in that area would overlook.

We know this is not the most efficient way to run a laboratory. We know it will not do for a crash program. But when you can do it, *do* it. In the long run, you will build breadth and flexibility into your organization, because you will be building diversified skills. And these will open the door for creative advances all along the line.

Donald C. Pelz
and Frank M. Andrews

6 Solving Problems

Psychologists have been experimenting on thinking for more than sixty years. They have observed the problem-solving behavior of chickens, rats, chimpanzees, children, adults, mental defectives, psychotics, and so on. They have gathered data on how rats release themselves from puzzle boxes, how chimps employ tools to win bananas, how children combine colorless liquids to find the combination that produces yellow, and so on.

Is there anything that can be said, on the basis of this research, to help people solve practical and technical problems? When we — both of us experimental psychologists — first asked ourselves this question, we agreed the answer was "No." It seemed to us that such studies were too far from naturalistic settings to have relevance for problem-solving in "real life." But then we looked again at the outcomes of a number of experiments — and we were pleasantly surprised. It seems to us now that there may be a common thread between many of the laboratory experiments and many kinds of problem difficulties that crop up in our daily activities.

When we stand back and survey this heterogeneous body of data, we see one theme that appears again and again. In attempting to solve a given problem, the most typical difficulty is that humans fail to make use of the information *they have.* Since most of this article will deal with laboratory experiments that are concerned with this difficulty, it may help at this point if we give two examples from "real life."

The first must be as old as the inner tube: It's the homely

story of the truck that was stuck in an underpass. Various onlookers tried to be helpful by suggesting ways for extricating the truck, but all these suggestions involved reasonably major deformations — either of the truck or the underpass. Then a little boy suggested letting air out of the tires. Many such stories exist in science and invention. All serve to show the same point: A solution, once stated, can be seen as "obvious."

A classic example involves the invention of the ophthalmoscope. The physiologist Brücke, interested in how the retina of the eye reflects light, devised an instrument to illuminate it. The famous physiologist Helmholtz, while preparing a lecture on Brücke's device, suddenly realized that the rays reflected from the illuminated retina could be used to view the retina itself. Helmholtz used a series of mirrors to reflect light into and out of the eye, and then used a lens to form an image with the light reflected from the retina. This was, as the mathematician Jacques Hadamard points out, "an almost obvious idea, which, as it seems, Brücke could hardly have overlooked." But overlook it he did. Why? Here is Helmholtz' explanation: "In this, Brücke was within a hairsbreadth of the invention of the ophthalmoscope. He did not ask himself what optical images are produced by the rays that emerge from an eye into which a light is thrown. For his particular purposes it was not necessary to ask this question, but had it been posed, he would have been able to reply to it as quickly as I have."

Nobody should be surprised when people fail to solve problems for which they lack the necessary information. But it is curious indeed that they fail so frequently when they have all the necessary information.

How is it possible for a person to have the necessary information and not be able to use it? The answer seems to lie in the fact that the brain, like the computer, is divided into a storage unit and a processing unit. While the storage unit can hold a vast amount of information, the capacity of the processing unit is strictly limited. The average person is able to retain and repeat back immediately only about seven unrelated digits. This fact, along with other research, suggests that the processing unit can handle no more than about seven independent items of information at a time. Now any problem of any consequence probably involves more elements than can be handled by the immediate memory span. Unless the processing unit is guided by a systematic search plan, and unless it possesses perfect memory of where it has already looked, it can easily overlook elements or combinations for consideration. Furthermore, not only must the processing unit sift the relevant from the irrelevant — it must also organize the elements into larger units in order to deal with more items in its attempt to construct a workable solution pattern.

The limited attention span and the necessity of "chunking" items of information into larger organization units may contribute to an individual's failure to use available information. He may start his search by looking at the wrong elements. This will not necessarily lead him into trouble. But as he looks at any set of elements he is simultaneously placing them within a tentative organization. And this initial organization — no matter how tentative — serves as a guide to his search process. If it is not an appropriate construction, this organization may prevent him from looking in more appropriate directions. For example, think again of the truck: An inappropriate construction directs one's attention to the top of the truck, since this is where the problem is. And, if your thoughts are channeled by the inappropriate construction, this is where you look for the solution.

How can the problem-solver be helped to look in the right direction? In one sense, we cannot answer the question, for we do not know in advance what the right direction is. But in another sense we can answer, by pointing out that the problem-solver is more likely to hit upon the correct approach if he tries several approaches. What so frequently produces failure in problem-solving is getting stuck on one approach and being unable to abandon it.

Let us look now at several precepts — or working rules — that may help in the problem-solving process. These precepts fall into two categories: those intended to keep the problem-solver from getting stuck on an incorrect line of attack and those that may be expected to help him get free when he is stuck. In other words, preventive rules and remedial rules.

Four considerations guided our choice of precepts. One, we agreed they must be relevant to overcoming the difficulty people have in making use of information. Two, they must be operational; that is, they should specify concrete actions that the individual can take. Three, they must be applicable at the time the individual finds himself confronted with a problem. (This forced us to exclude many precepts that might be classed as "preparation for problem-solving" or "education for problem-solving.") And four, they must have some support in psychological research.

Precept I: Run over the elements of the problem in rapid succession several times, until a pattern emerges that encompasses all these elements simultaneously.

This precept helps keep you from prematurely fixating upon a subset of the elements required for the solution. It also helps you get the "total picture" before you become lost in the details.

Descartes includes this precept in his *Rules for the Direction of the Mind.* He describes its application as follows: "If I have first found out by separate mental operations what the relation is between magnitudes A and B, then that between B and C, between C and D, and finally between D and E, that does not entail my seeing what the relation is between A and

E, nor can the truths previously learned give a precise knowledge of it unless I recall them all. To remedy this, I would run them over from time to time, keeping the imagination moving continuously in such a way that while it is intuitively perceiving each fact it simultaneously passes on to the next; and this I would do until I had learned to pass from the first to the last so quickly that no stage in the process was left to the care of memory, but I seemed to have the whole in intuition before me at the same time. This method will relieve the memory, diminish the sluggishness of our thinking and definitely enlarge our mental capacity."

Compare this quotation with that of Helmholtz, some 200 years later: "It is always necessary, first of all, that I should have turned my problem over on all sides to such an extent that I had all its angles and complexities 'in my head' and could run through them freely without writing."

Precept II: Suspend judgment. Don't jump to conclusions.

This precept serves to keep you from getting stuck on the first one or two interpretations that come to mind. When we jump too quickly from the problem statement to an attempted solution, we frequently get trapped into clinging to an inappropriate direction.

Two examples: In teaching sixth-graders how to ask the right kinds of questions in order to discover the scientific principle underlying a physical event, Suchman found it important to train the children from prematurely guessing at the explanation, for once they had offered an explanation, these children had difficulty revising it or dropping it in the face of contradictory evidence. Bruner and Potter show this in another context: Their experiments show the fixating power of premature judgments. Color slides of familiar objects, such as a fire hydrant, are projected upon a screen and people try to identify the objects while they are still out of focus. Gradually, the focus is improved, through several stages. The striking finding is this: If an individual wrongly identifies an object while it is far out of focus, he frequently still cannot identify it correctly when it is brought sufficiently into focus so that another person — who has not seen the blurred vision — can easily identify it. What this seems to say is this: More evidence is required to overcome an incorrect hypothesis than to establish a correct one. He who jumps to conclusions is less sensitive to new information.

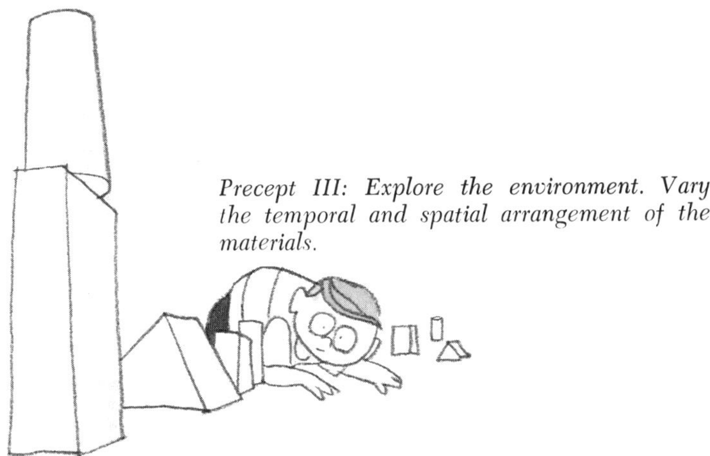

Precept III: Explore the environment. Vary the temporal and spatial arrangement of the materials.

This precept serves to keep the mind "loose" by activating a variety of possibilities. It may also help uncover familiar patterns that were masked by an originally unfamiliar arrangement.

The difficulty in many problems frequently resides in the way the elements happen to be ordered. A rearrangement of the elements sometimes reduces the problem to a task that can be handled by standard procedures. The Scotch psychologist Hunter, for example, found his subjects had difficulty specifying the relation of George to Willie when he gave them these two terms: Harry is shorter than George; Harry is taller than Willie. Many eleven-year-old children cannot handle this task, and the adult subjects took considerably more time to give their answers than when the problem was put in its logically equivalent (but more familiar) form: George is taller than Harry; Harry is taller than Willie. In fact, the evidence indicates that subjects actually solve this problem by rearranging the elements until they correspond with the familiar form: A is greater than B; B is greater than C.

Quite frequently, the solution to a difficult problem is suggested to the solver by a slight change in his physical relation to the elements in the problem. In Köhler's famous experiments on chimpanzees, one chimp was faced with the task of obtaining a banana that was beyond arm's reach through the bars of his cage. A stick was available to him, but the stick was behind him and out of his field of vision when he looked at the banana. Later, when he was idly playing with the stick, the banana and the stick accidentally became part of the same visual field. When this happened, the chimp instantly made the connection between the stick as a tool for extending his reach and the obtaining of the banana.

Now we shall turn from "preventive" precepts to others that can be looked upon as special operations for accomplishing the following more general precept: If you are getting nowhere on a problem, abandon your approach and try to find a new difficulty as a basis for solving the problem.

Let us begin with the word "direction" — a word employed by Maier to refer to the way in which an individual will attempt to solve a problem. The direction he takes, Maier tells us, depends on what he sees the problem to be. An example: In one of his experiments, Maier gives his subject the task of tying together the ends of two strings that are suspended from the ceiling; the strings are located in such a way that the subject cannot reach one string with his outstretched hand while holding the second string in his other hand. The typical person will see the difficulty as a shortness of reach. Consequently, his "direction" will be toward ways of lengthening his reach — by searching for a stick or hook, for instance. Another person will see the difficulty as a shortness of one of the strings. And consequently, this person will try somehow to make one of the strings longer. Now Maier arranged things so that these obvious solutions could not be used; he wanted to discover some things about "good" reasoners and "poor" reasoners, so he devised his test in a way that required a more imaginative solution: The solution he was looking for required that the subject see the difficulty in terms of getting the second string to come to him. Quite simply, if the subject tied something to the end of that string and then caused it to swing as a pendulum, so as to be grabbed as it swung toward the subject while he held the first string, then the subject would have solved the problem correctly.

The insight Maier gained from this test is this: He found that good reasoners do not persist in one direction if they are getting nowhere. Rather, the good reasoner will jump from one direction to another until he finds a solution. Poor reasoners, on the other hand, persist doggedly in the same direction, even when the difficulty does not yield to their efforts. But he wanted to know more than this. For instance, as he watched people fumble along in the wrong direction he wondered whether this simply indicated that such people were incapable of better reasoning, or whether they were being blocked from considering new directions by some stubborn commitment to the old. To provide himself with an answer to this question, Maier performed another experiment. If people were being blocked as he suspected, he reasoned that such people could reach solutions sooner if they were warned against continuing with an unsuccessful experiment. On the other hand, if such people simply were incapable of better reasoning, then such warnings would make little difference, since they would not be able to devise better alternatives anyway. Maier's experiment consisted of giving several hundred college students a one-hour test. Each was asked to solve three problems — one being the two-string problem of a moment ago. The students could do the problems in any order, and they could divide their time among the problems in any way they wished. With half the students, Maier provided the problems with no preliminary comments. With the other group, he prefaced the test with a brief introductory lecture, which concluded with the following hints:

1. Locate a difficulty and try to overcome it. If you fail, get it completely out of your mind and seek an entirely different difficulty.

2. Don't be a creature of habit. Don't stay in a rut. Keep your mind open for new meanings.

3. The solution pattern appears suddenly. You cannot force it. Keep your mind open for new combinations and do not waste time on unsuccessful attempts.

As one might expect, this brief lecture produced a significant improvement in performance. The male students who heard the lecture, for example, solved an average of 62 percent of their problems, while those men who heard no preliminary comments solved only 51 percent of theirs. Among the young women, the same pattern appeared: 36 percent for those who heard the lecture, compared with 25 percent for those who did not.

From these tests and others, Maier concluded that the persistent and initially wrong directions that accompany problem-solving actually prevent correct solutions from appearing. "Reasoning," he said, "at least in part, is the overcoming or inhibiting of habitual responses."

This precept serves to shift the problem-solver's orientation from solution-mindedness to problem-mindedness. In their studies of problem-solving, Maier and Hoffman found that, when an individual was seeking his first solution to a problem, he was dominated by strong pressures to achieve that solution — in other words, he was solution-minded. When he was encouraged — after the first solution had been achieved — to go on to seek a second solution, the second solution was usually a more creative one. This was so, they concluded, because the individual was problem-oriented — he was no longer driven to find *a* solution, for he had already accomplished that; now he was, as Helmholtz would have said, "turning the problem over on all sides." Maier and Hoffman found this precept to increase "creative" solutions from 16 percent to 52 percent.

Precept IV: Produce a second solution after the first.

I want to insert a personal observation here — this is Hyman talking now: I have found that, unless subjects are told in advance that they will have to produce a second solution, that second solution may be inferior to the first. But I do want to underline the importance of the Maier-Hoffman discovery by stating it again, in a somewhat different form: We can make much better use of the information we have at hand when we are *pushed to the limit.* An example from my own research: I asked a group of people to produce three solutions to the same problem. The problem related to the declining teacher-student ratio in higher education. I gave them statistics and information relevant to the problem and asked each person to write his first solution to the problem of maintaining high-quality education in the face of this declining ratio. Then I asked them to perform some other tasks, and after this interval they were asked again to reconsider the educational problem and write their second solutions. And after another interval I asked them again to produce their third solutions — but on this third go-round I deliberately blocked them from using ideas they were familiar with. I did this by giving each person a copy of his first two solutions, as well as a list of some of the most commonly offered solutions. And I said that these solutions, being already known, should not be considered for solution number three. What we wanted now were solutions that did not make use of any of these ideas. At this point, about one-quarter of the participants threw up their hands in despair. They could think of nothing to do or say. But the others! All were able to come up with at last adequate solutions — involving completely new ideas, in some cases to their own surprise. And as many as 25 percent of these people came up with truly creative and outstanding answers. They had been blocked from using obvious answers — they had been pushed to the limit.

Precept V: Critically evaluate your own ideas. Constructively evaluate those of others.

This precept provides a heuristic for discovering new directions. Its purpose is to guard you against complacency in accepting your own ideas for solving a problem — and also, to help you get new ideas and insights from the attempts others have made at solving the same problem. Torrance, working with students, found that he could increase the creativity of term papers — in which students devised an original research project — by inducing attitudes of constructive evaluation of others' ideas.

I (Hyman) tested this precept with a group of 36 engineers. I asked them, individually, to devise a solution to an automatic warehousing problem. With half the group, I provided some solutions that other engineers had proposed — then I asked these people to list reasons why those solutions could *not* work. (This is the kind of critical task engineers seem well suited for. They willingly accepted it.) Following this task of critical evaluation, I asked each of these 18 engineers to write out his own solution to the problem. With the second group, I showed the same set of previous solutions, but I asked these men to list as many *strong* points as they could find within those previous solutions. Then, following this constructive evaluation task, these engineers wrote down their own solutions. My next step was to have these 36 solutions evaluated. I mixed them up, thus to prevent anyone's knowing the origin of any particular solution, and gave all 36 solutions to a committee of experts, who rated them for creativity. When the evaluations had been made, it turned out that the solutions of those engineers who had been asked to find strong points were significantly better — more creative — than those who had been asked to look critically at the solutions of others.

Before we can really pin down the extent of, and reasons for, these effects, we must have more information. But one thing that is suggested here is that a positive attitude — or a negative attitude — toward one's own ideas and toward the ideas of others can markedly affect the quality of the solution to a problem. More important, a constructive attitude can easily be induced by the simple expedient of causing the individual to look for weak points in his own first ideas and for strong points in the ideas of others.

Precept VI: When stuck, change your representational system. If a concrete representation isn't working, try an abstract one, and vice versa.

This precept takes advantage of the fact that relationships that are not easy to see in one representational system are often obvious in another. If you have been dealing with a problem in verbal terms, and if you are stuck, then try to switch to a picture, a model, a graph, numbers, or even to other words. If you have been dealing with the problem in nonverbal terms, try mapping the elements into words. A study, by Mawardi, of a group of professional creative thinkers bears this out. She taped sessions in which the group worked on a single problem. Then she classified these sessions into idea units — some were classified as abstract, others as instrumental, metaphor, orientation. She found a very strong tendency within the group to alternate between abstract and concrete modes of thinking.

Precept VII: Take a break when you are stuck.

This precept takes advantage of the fact that a dominant direction will usually weaken with time. Now, surely, to take a break when stuck is the most frequently given advice to problem-solvers. But does it really do any good? In two experiments at Stanford, the evidence seemed contradictory: Irvine and Taylor found no advantage in taking a break, while Taylor and LaBerge — in the second experiment — found a definite advantage. The difference was this: There was an advantage when the subjects were allowed to determine their own times for taking breaks. But what is the proper time? And with this question, we must face up to still another question: What is the meaning of "being stuck"? Let us look at the situation in this way: You are getting nowhere with a problem. Why? It may be that your plan of attack is inappropriate. Or it may be that your plan of attack is OK but your perception of the materials is such that they cannot be adapted to your plan. In either case, it is doubtful that a break will help you, for it may only cause you to return to an inappropriate plan or perception. On the other hand, if you have explored the possibilities of your present approach rather thoroughly — and if you cannot think of another approach to try — then you are "stuck" and this would seem a good time to take a break. Put it this way: If you are really stuck, then take a break. But if you simply have not given an approach sufficient thought, then the break won't help.

In talking with someone, you are forced to consider aspects of your problem that you might not otherwise have considered. You cannot take "short cuts" here; you cannot jump across gaps that you might otherwise take for granted. You must return to fundamentals in order to communicate with your listener. And the presence of the listener provides a powerful feedback mechanism that quickly detects obscure or inconsistent points in your story. Where you have to communicate to another person, you have to put yourself in his shoes; this in itself is a powerful precept.

Precept VIII: Talk about your problem with someone.

Further, the act of communicating any problem transforms that problem from a private to a public form. The Swiss psychologist Jean Piaget places great emphasis on this role of communication in the development of the thought process. Piaget has made monumental investigations of how thinking develops in children, and he has concluded that thinking—as we recognize it in our society—would never develop in children if they were not forced by society to justify with reasons their behavior to adults and peers. (Quite independently, Russian psychologists have come to the same conclusion.) By being forced repeatedly to communicate his ideas to others, the child gradually masters public forms of representing the world. As he grows older, the child internalizes this public system of representation and thus can check his ideas by recasting them in this public form. For many kinds of thinking, he can now operate independently of the actual presence of others.

But for especially difficult problems, it still helps to communicate with someone. The importance of this precept is illustrated by an incident from contemporary medical research. Dr. Lewis Thomas, in a study involving the enzyme papain, noted a striking phenomenon when he injected the enzyme into the blood stream of a rabbit: The rabbit's ears wilted dramatically. Thomas immediately interrupted his original investigation to try to discover why. He sliced and stained the rabbit's ears. He saw nothing unusual in the connective tissue — where he expected to find changes. He also looked at the cartilage, but he saw nothing obvious there. Like all physicians a few years ago, Thomas had looked upon the cartilage as inactive — he certainly expected to find no changes there. After searching in vain to solve the case of the floppy-eared rabbits, he gave up and returned to his regular work.

About seven years later, Dr. Thomas was teaching second-year medical students how to perform laboratory studies. He decided to demonstrate the floppy-ears phenomenon, feeling it would capture the students' interests. He went through the standard tests again, searching for the cause of the phenomenon. And this time he found it — in the cartilage. What made the difference? Here he was communicating to students; he did not skip or overlook incidental steps and precautions. In his own words: "Well, this time I did what I hadn't done before. I simultaneously cut sections of the ears of rabbits after I'd given them papain *and* sections of normal ears. This is the part of the story that I'm most ashamed of — the only way you could make sense of this change was simultaneously to compare sections taken from the ears of rabbits that had been injected with papain with comparable sections from the ears of rabbits of the same age and size that had not received papain."

Here we see, in striking fashion, how the necessity to rearrange his thinking led the investigator to take precautions he would normally overlook. The result was a discovery that he could have made years earlier.

By phrasing them in a very general form, our eight precepts can be reduced to two:

Look before you leap.

After you have leaped, if you find yourself bogged down, find out what you are doing and then do something else.

The usefulness of these precepts depends upon the manner in which they will be applied. And it also depends upon future research — psychological research. But let us add this much for psychological research: As a result of writing this paper, we have discovered that psychological research has more to tell about problem-solving than even we had realized. And further — and this may testify to the fruitfulness of the precepts — in trying to communicate what we know to people of the "hard" sciences, we have come up with a number of ideas for doing experiments on problem-solving.

Ray Hyman and Barry Anderson

The Environment Part II

7

The Growth of Ideas

Throughout these next two parts of the book, you are going to encounter a pair of related activities. One is invention and the other is innovation. Each demands a high level of creativity. Each involves the skills of technical specialists. But you must not confuse one with the other, for the two are separate but essential components of the innovation process. Invention is perhaps more fun to read about, because it is a subject charged with drama, with visions of the starving inventor in his lonely attic, then his triumphant discovery and, ultimately, the acclaim of a grateful world. Further, it is drama whose characters are complex and often fascinating individuals. As historian Elting Morison says of some 30 inventors who flourished in the last century: "A surprising number turned out to be people with little formal education, who drank a good deal, who were careless with money, and who had trouble with wives or other women." But if invention is the dramatic stuff of this narration, innovation is the fire that makes it bubble.

I suppose Morison's characterization holds for many inventive men of our contemporary world, but the thing that makes this such an interesting subject — indeed, the reason it is such an important area for exploration — is the fact that invention and innovation often fail to get together. What happened? What went wrong? Why did we waste all that money on a bad idea? Actually, there is still another thing that gives the subject interest and amplifies those questions when the innovation process fails: When invention and innovation *do* get together, when the process works as it is supposed to, things really happen. And they happen fast. Instead of existing in an industrial environment with a 40-year cycle of innovation, you may find yourself running at four times that speed, because your industry may now go through a technological revolution every ten years. This, too, is the stuff of which many yarns are spun, but it is also the stuff that can drive you crazy.

In these next essays, we are going to talk about this gaming parlor called the research environment, and we will have some things to say about the habits of the gamblers, be they the research managers or their bosses, the top managers. As you read these pieces, you will find some general principles for a productive technical environment. For example, you will see the importance of involving the technical people themselves in helping to define their organizations' goals. And you will see the crucial role of the person we call "the champion," that zealot who will bull the idea through, whatever the hazards. We heard earlier of William Shockley and how he played this role in the development of the transistor, but we might also have cited Robert Goddard and rocketry, or Hyman Rickover and nuclear-powered ships, or Frank Whittle and jet engines, or Edward Teller and the hydrogen bomb, for each of these men played the champion role. But you will find no simple formula for specific successful discovery, nor will these essays provide you with an infallible way to distinguish the zealot who is right from his numerous brethren who are just as determined and just as convincing but whose ideas turn out to be wrong or irrelevant. It is inspiring to hear how the work of one man can provide the base upon which others may build, but such success stories, however informative and inspirational, really tell us only part of the story — the successful part.

The point that must be made, therefore, before we go on to discuss the principles that relate to the creation of a productive technical environment, is that there is no simple way to describe the very complex process by which ideas are caused to occur and develop. Nor is there a technique by which to measure the potential value of a new idea, or the so-called cost-effectiveness of a particular technical environment. This is not to say that the general principles are useless, but rather that they are neither detailed nor quantitative. For example, it is generally true that basic reseach by itself rarely provides a new and unexpected idea that inspires people in engineering to try to develop the idea into a useful new product or process. There will be people in basic research who will challenge this statement, but there is much historic evidence showing that it *is generally true.* And yet it is not *always* true, and that is the important thing to remember: Sometimes important technical developments *do* violate this general principle, as was the case with the discovery of polysulfide polymers, which ultimately resulted in the spectacular success of the Thiokol Chemical Company, and as with the transistor, whose origin was the Bell Telephone Laboratories.

This takes us close to a point that Donald Schon discusses in his essay on innovation: that a corporation "is precisely an organization designed to uncover, analyze, evaluate, and operate on risks." Schon maintains that a corporation cannot operate in uncertainty, but that it is "beautifully equipped" to handle risk. How is this done? Says Schon: the work begins with more information than can be handled; this information is then operated on, at lower levels of the corporation, until clear alternatives of action, together with their probable benefits and risks, can be defined. At this point, management can play the investment game — the game of deciding where to put its bets. Supplementing Schon's observation, Albert Stewart points out in his essay the importance of clear communication between the industrial scientist and management, so that the gamble is indeed being taken on the idea that management *believes* it is financing. What all of this adds up to is the following: Certainly invention is a risk-taking venture, and certainly the innovation process is a delicate process, but to the manager who understands this process it is not the unmanageable, dangerous, bewildering thing it appears to be to the manager who does not understand it. Schon says a corporation's research and development activities do not enjoy the same degree of predictability as do other parts of the organization, such as production and accounting, and he warns that innovation is made impossible when management insists upon such predictability. I said in the introduction that the authors of these essays do not necessarily agree with all of the observations of their fellow authors, and we have here a case in point. You will discover this when you have compared Schon's comments with those of Jack Morton, who fears you may conclude from Schon that innovation is far riskier than it actually is. "The risk of *not* doing innovation is the larger risk," Morton says, "and the risk of doing irrelevant research is larger than doing relevant research."

When you have read these essays and reflected upon their various shades, I believe you will have a clearer picture of that condition I described earlier, the productive technical environment. Schon, Morton, and others describe it in terms of innovation, the process of introducing new technology to use. Carl Barnes talks of it in terms of productivity, particularly the productivity of large industrial laboratories. With the Schon and Barnes pieces, we want to provide what might be called a physiological view of the technical environment: Why is innovation feared? What can be done to create a healthier environment for invention? Then we shall go into the laboratory itself, with Lowell Steele's two pieces on the technical manager: the man in the middle, the man who must try to reconcile the needs and aspirations of the scientist and the mission of the scientist's organization.

With Steele's two essays, you may well conclude that the task of resolving such divergent objectives — the scientist's versus his organization's — is impossible. You may conclude that industrial science cannot work. But such conclusions will be too harsh, for despite its difficulties, frustrations, and embarrassing failures, the idea that began to take form so many years ago — with Bell in Boston, Edison at Menlo Park, and Whitney at Schenectady — has turned out to be a very good idea. Perhaps Alfred North Whitehead was correct when he described the *method of invention* as "the greatest invention of the last part of the nineteenth century." A distinguished committee of the National Academy of Sciences looked not long ago at the world of American industrial research and concluded it had been a "magnificent triumph."

You will encounter a phrase later on that describes the environment in which many accomplishments of industrial science have occurred: "participative management." The phrase was coined by Rensis Likert, who winds up our section on the technical environment. Scientists and engineers are likely to be most productive, Likert declares, when their supervision is such that they feel substantial freedom in their work — in selecting their problems and goals, in deciding on the approach to achievement, and in interpreting their data — and when they have frequent interaction with their superiors. This does not mean that technical people are most productive when supervision is simply well intended, nor is there likely to be a productive technical environment if the technical people alone are the setters of goals. As Steele says in describing the role of the manager, the technical man is vitally interested in his boss's competence to render *knowledgeable, balanced judgments.* He finds it rather unsettling to have to lead his boss step by step through his work, to have him ask questions that indicate that he doesn't really understand what is going on. What is even worse is to have the boss get enthusiastic about what the man regards as "the wrong things."

The National Academy of Sciences has recently contributed to our knowledge of the innovation process through a pair of studies of the technical environment:

One study was conducted by the Academy's Panel on Applied Science and Technological Programs. Harvey Brooks of Harvard was chairman of this group, which included technical leaders from such industrial organizations as Union Carbide, Ford Motor, General Electric, and Bell Telephone Laboratories, as well as a few nonindustrialists, including Edward Teller. It was this same group, by the way, that said that American industrial research had been a magnificent triumph. Among its many conclusions and recommendations regarding the nature and strategy of applied research, the first, in part, is the following: "In examining examples of successful translation of science into technology, one is struck by the diversity of successful patterns and organizational structures. There are no simple formulas for success, and for this reason success is most likely when laboratory management has wide latitude in adapting and restructuring the organization to suit the particular problem areas or technologies with which it currently dealing." Regarding the matter of participation of the scientists themselves in setting the organization's goals and strategies: "The key individuals in the research organization are fully aware of and sympathetic to the principal goals of the organization. . . . The organization has a quick response in recognizing and funding new ideas. . . . At each organizational level, the individual responsible has some freedom in redeploying the resources at his disposal without extensive review by higher authority."

The second study dealt with recent accomplishments in materials science and technology, e.g., the development of silicones, Pyroceram glass, and superconducting magnets. Virtually all of these investigators were industrial scientists and engineers, e.g., from Koppers, Arthur D. Little, and Corning Glass, under the chairmanship of Morris Tanenbaum, the director of research and development of the Western Electric Company. They too found that freedom and success were frequent partners; in fact, more often than not it turned out to be *critical* that the research people be able to shift the directions of their work and explore unanticipated but relevant paths, and that such shifts be made at the discretion of the technical people themselves, without waiting for review and approval by top management.

Lest you read the above and the pieces that follow with an undue sense of self-aggrandizement, I suggest you take special note of a personal inquiry, which appears at the conclusion of these introductory remarks. The inquiry was developed by Tanenbaum and his colleagues, and the questions turned out to be valuable to them in their effort to define and understand the circumstances surrounding the technical developments they investigated. We present the questions for your consideration because your answers — along with the answers you receive from others in your organization, including nonsupervisory technical people — may reveal some new knowledge of your own organization's effectiveness. We suggest you come back to these pages after you have read the next two parts of the book.

David Allison

Ask Yourself These Questions

1. How much flexibility is available in level and allocation of funding and support to the members of the technical staff reporting to you? How much flexibility do you have? How much flexibility do your contractors have?

(a) What approvals are needed to change the level and allocation of funding?
(b) What are the mechanisms for changing the level and allocation? How complex are the mechanisms? What time interval is usually required?
(c) What is the effect of your organization's structure and controls on these mechanisms? Do they inhibit or stimulate change?
(d) What is the balance in your organization between flexibility and the need for control?
(e) What are the penalties and rewards involved for those who make the sometimes risky decisions to change the levels of funding and support?

2. How much flexibility is available to those reporting to you with respect to changing directions and goals?

(a) What approvals are needed to change directions and goals?
(b) What are the mechanisms for changing directions and goals? What time intervals are usually required?
(c) What is the effect of your organization's structure and controls on these mechanisms?
(d) What is the balance between flexibility in this area and the need for control?
(e) What are the penalties and rewards involved?

3. How much flexibility is available to those reporting to you in starting (initiating, funding, and actually beginning work on) new projects or terminating existing projects?

(a) What approvals are needed to start new projects or terminate existing ones?
(b) What are the mechanisms? What time intervals are usually required?
(c) What is the effect of your organization's structure and controls on these mechanisms?
(d) What is the balance between flexibility in this area and the need for control?
(e) What are the penalties and rewards involved for those who make the sometimes risky decisions to initiate or terminate projects?

4. Does each individual in your organization clearly understand the flexibility and limits described above that are available to him and the mechanisms, balances, penalties, and rewards that are involved?

5. How does the distribution of technical effort (i.e., in-house, contract, subcontract) affect your flexibility? How can the balance of distribution be changed?

(a) What approvals are needed to change the balance?
(b) What are the mechanisms for changing the balance? What time intervals are usually required?
(c) What is the effect of your organization's structure and controls on these mechanisms?
(d) What is the balance between flexibility in changing distribution and the need for control?
(e) What are the penalties and rewards involved for those who make the sometimes risky decision to change the distribution of effort?

6. How far can those reporting to you carry their work through the research-development stages? What are the restrictions on the directions in which they can work?

(a) To what degree does the formal mission of your organization affect the stages and directions in which you can work?
(b) What are the mechanisms by which you can overcome the limitations?
(c) How can you influence other groups to carry your work forward? To support your work?

7. Are frequent communications between your group and other groups needed? What are the specific groups with which such communications are needed? In which of the research-development stages do they work?

8. What formal mechanisms exist to assure adequate communications betwen your group and others?

9. Can you identify "informal" or "unofficial" structures or mechanisms that serve to promote communication between your group and others? What are they?

10. What barriers (organizational: such social aspects as status differences, cliques, military-civilian funding, etc.) exist that may inhibit communications between your group and others?

11. What has been done in your organization to enhance communications?

12. Who are the individuals in your organization who have acted as couplers?

(a) How do organizational factors affect the way these people function as couplers?
(b) Do you give specific recognition to the coupling function?

13. Where and who are the people in your organization upon whom you rely to accomplish the transition between research findings and engineering practice?

(a) How do your applied research people communicate with people working in relevant areas of fundamental research and with people who have end-item responsibilities?
(b) How do your basic research people stay abreast of relevant applied research?
(c) How do those with end-item responsibilities stay abreast of relevant applied and fundamental research?

14. Who are the people who have been "champions" in your organization?

(a) What kinds of barriers did they have to overcome? Do the barriers still exist?
(b) How was the "champion" rewarded? What is his role today?
(c) What effect did he have on the organization?

15. Are the mission, goals, and long-range objectives of your organization clearly defined?

(a) Are they defined by a technical or administrative part of the organization?
(b) What specific role do you play in defining the mission, goals and long-range objectives? What role do the people who work for you play?
(c) Do the people who report to you have a full understanding and perspective of the mission, goals, and long-range objectives? How is this achieved? Upon whom do you depend for this understanding and perspective?
(d) How much of the work performed by your organization clearly conforms to the defined mission, goals, and long-range objectives? How much deviation, if any, is permitted? Who decides?

16. Does your organization have a commitment to the desirability or necessity for research-engineering interaction? If this commitment exists, what do you do to make it visible?

17. With what frequency do individuals transfer between basic-research groups, applied-research groups, and engineering groups?

(a) Do transfers occur in both directions?
(b) Do transfers occur between line and staff organizations?

18. What is the distribution of formal education in the various parts of your organization?

(a) What formal educational background is required for the assigned responsibilities of the various parts of your organization?
(b) How does it match between groups whom you expect to communicate and interact with each other?
(c) Does the distribution foster or impede communication?

19. Who defines the specific needs to which your organization is expected to respond?

(a) Are the needs defined by a technical or administrative part of the organization?

(b) What specific role do you play in defining the needs? What role do the people who work for you play?

20. How well specified are the needs to which your organization is expected to respond?

21. How are needs communicated to you? Are there formal mechanisms for presenting needs?

22. In those instances where you have used the findings of basic research to help satisfy a current need, when was the pertinent basic research performed?

(a) If the research was relatively old, why do you think it had not been applied earlier?

(b) What was the mechanism by which it was now applied?

The Fear of Innovation 8

There is a "rational view" of innovation. According to this view, innovation is similar to other major functions of an organization — such as sales, accounting, or production. Therefore, if we accept the rational view, we say that innovation is a *manageable process* in which risks are controlled by mechanisms of justification and review.

I want to show that the rational view of innovation ignores or violates actual experience. In light of that experience, the notion of innovation as an orderly, goal-directed, risk-reducing process must appear as a myth. I shall explore some of the ways in which it is a myth, then go on to show that when an organization lives by the myth it can discover that it has made innovation impossible.

Implicit in the rational view is the notion that skilled men can anticipate and control the risks of innovation. Phrases like "the management of innovation" suggest that we can foresee and quantify the likely dangers and rewards of a technical project and weigh them against the likely risks and rewards of alternative efforts. By selecting only those projects whose benefits justify their anticipated costs, by playing risks off against one another — in short, by a process of justification, decision, and optimization — we can (it is assumed) keep the risk of innovation within bounds.

Risk has its place in a calculus of probabilities. It lends itself to quantitative expression — as when we say that the chances of finding a defective part in a batch are 2 out of 100. In the framework of benefit-cost analysis, the risk of an innovation is how much we stand to lose if we fail, multiplied by the probability of failure.

Uncertainty is quite another matter: A situation is uncertain when it requires action but resists analysis of risks. For example, a gambler takes a risk in an honest game of blackjack when, knowing the odds, he calls for another card. But the same gambler, unsure of the odds, or unsure of the honesty of the game, is in a situation of uncertainty.

How does this apply to technical innovation? Men involved in technical innovation in a corporation confront a situation in which the need for action is clear but in which it is by no means clear what to do. This situation is painful and full of anxiety for the individuals and, in a sense, for the corporation as a whole. To be in a corporation in a time of corporate uncertainty is to be engulfed in waves of anxiety. So long as this situation obtains, the corporation cannot function effectively. The corporation is not designed for uncertainty — where there are no clear objectives to reach, no measures of accomplishment, and where it is not clear what to try to control. A corporation cannot operate in uncertainty, but it is beautifully equipped to handle risk. It is precisely an organization designed to uncover, analyze, evaluate, and operate on risks. Accordingly, *the innovative work of a corporation consists in converting uncertainty to risk.*

The work begins with more information than can be handled; this information is then operated on, at lower levels of the corporation, until clear alternatives of action, together with their probable benefits and risks, can be defined. At this point, management can play the investment game — the game of deciding where to put its bets. The game requires analysis of investment alternatives, estimating their markets, costs of technical feasibility, and making investment decisions. The game is played with competitive corporations as opponents. The rewards and punishments of the game can be measured in dollars.

In the process of innovation, everything is done to permit decision on the basis of probable dollar costs and dollar benefits. In the process, the corporation converts the language of invention to the language of investment. Instead of talking about materials, properties, performances, experiences, experiments, and phenomena, the corporation talks of costs, shares of market, investment, cash flow, and dollar return.

The conversion of uncertainty to risk takes varying forms, depending on the *kind* of uncertainty to be dealt with. Technical innovation involves many different kinds of uncertainty. Some of these spring directly from the nonrational character of the process of invention; others are only indirectly related to that process. I will be concerned here with the uncertainties springing from determination of technical feasibility, novelty, and market.

One focus of uncertainty is in the question: Can it be done? Is it technically feasible?

In the process of invention, the requirements to be met change continually in response to unexpected findings. At any time, these may turn a good risk into a poor one — or an indifferent idea into an idea of great promise. We can describe these changes in terms of curves of technical difficulty. For example, as an investigator works over a period of time, he may first encounter his most difficult problem and later on only minor ones whose solutions he knows he can attain if he works long enough. If we plot his progress over time against "difficulty" or "demands of the solution with respect to the state of the art" the curve may look like this:

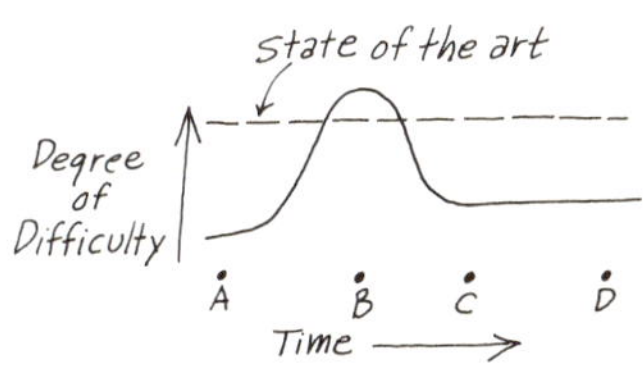

Here, the investigator must surpass the state of the art at *B*, whereas his subsequent tasks are well within the state of the art.

Or there might be a number of separate peaks. For example,

in considering the development of new semiconductor materials, we would have to identify separate peaks for the development of high-purity materials and for the development of high-yield production techniques. The curve might look like this:

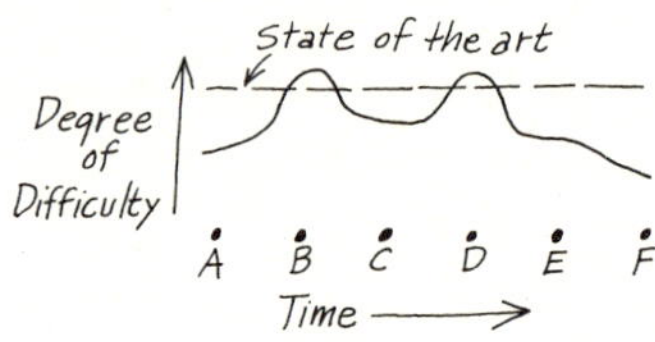

Such curves would not be a source of major uncertainty if it were possible always to identify ahead of time the problems and their degree of difficulty. Small improvements in the properties of materials — for example, raising by a few degrees the melting point of an alloy or increasing slightly the abrasion resistance of a plastic coating — may require no more than the continued application of known methods. On the other hand, achieving these improvements can turn out to be extremely difficult: A group working on the development of a new type of transformer, using piezoelectric principles, worked for weeks on what it thought was a materials problem, until it was able to make the calculations necessary to show that the performance requirements for which it was striving pressed to the limits both the energy capacity of the material and its ability to accept mechanical stress.

It is not always apparent, even to a skilled investigator, whether he is working on a minor problem of adjustment or a major problem of principle. He cannot place himself on the curve of technical difficulty. He may think he is here:

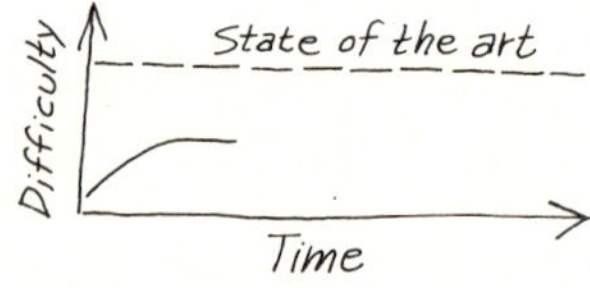

when in fact he is here:

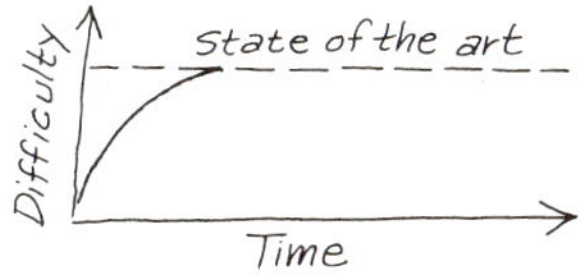

Technical feasibility, therefore, often resists the kind of definition required by the investment game. It may evade definition throughout the entire process of innovation.

Another focus of uncertainty concerns: Who has done it before? Who is doing it now? Although the patent literature offers a great deal of information about prior invention, a search may not uncover all relevant art. If a company undertakes a development, it can have no guarantee that its competitors are not doing similar things. Even if it knows that others are working in the same field, it cannot be sure how far they have gone. (The development of the automatic light meter for cameras is an example: The company that ultimately achieved commercial success had been spurred on by the belief that a competitor had the product, only to discover, later on, that its competitor did not have the product after all.)

And questions of marketing: Who will buy it? How large will its market be? How much of a share will it get? How long will it last?

The marketing questions have become the subject matter of a new profession — a profession that aspires to be a science. Without attacking the question of whether or not a marketing science is now possible — whether there are inherent uncertainties about marketing that no amount of information (acquired ahead of time) will resolve — we can see that the process of answering marketing questions occurs over a period of time and that there are bound to be stretches of uncertainty that can only be resolved by the expenditure of time and money, if indeed they can be resolved at all. The uncertainties of marketing are particularly apparent in the case of consumer products. Scotch tape, for example, was originally introduced as a way of mending books. Before its introduction, its makers had no inkling of the huge market potential it offered.

Companies make marketing decisions under pressure of money and time, confronted with more information than they can handle. Although one may be able to predict, in principle, the effects of single-variable changes — price, for example, or color — these variables never act alone. There are always many relevant variables, with multiple, interacting effects. There is no sure way of learning from past experience, for situations are never identical; differences that appear to be trivial may turn out to be critical.

Moreover, it costs money to find out. The introduction of a consumer product on a national scale is a major enterprise. A single regional market test, with appropriate preparation and follow-up, may cost hundreds of thousands of dollars — and yet the results may not permit formulation of the national picture. *In principle*, uncertainties may be resolvable, but the cost of resolution is high and the very process of resolving may cost more than can be justified.

To this we must add the fact that in the process of development there is interaction between need and technology: The market originally conceived for the product may be ruled out by technical limitations discovered in the process of invention. A different market may suggest itself — in this case, a more limited market. In short, the product for which a market must be anticipated is not an "it" that remains constant throughout the process of development. Rather, the product changes through this process — and its possible market changes with it.

Throughout the process of technical innovation, decisions about technical feasibility, novelty, and markets — among many other factors — must be made on insufficient evidence, by individuals who have more information than they can handle. The resolution of these uncertainties — the conversion of uncertainty to risk — takes time and money and requires justification in its own right. Its benefits must be balanced against its costs.

The process of innovation has a cost curve, as well as a curve of difficulty, which exhibits characteristic patterns. Let us consider the development of a new metal-to-metal adhesive. It begins in the laboratory when a chemist, working on another problem, notices an adhesive effect and reproduces it. Up to this point, the cost — in man-hours and materials — may be in the order of $5,000 to $10,000. The chemist shows the effect to the research director, who finds it intriguing and authorizes a search of the report and patent literature, some further experimentation, and a first examination of the markets. This may go on for a month or two and bring the total cost to $40,000.

At this point, the research director feels he has something worth presenting to management. Management likes the idea and launches a full-scale development project. The adhesive formulation has certain disadvantages, of course, and the development team makes efforts to improve it, as well as to explore variations of the chemistry involved, so as to "cover all the ground." A detailed analysis of the market — in the automotive, aircraft, and appliance industries — gets under way. It reveals the need for improved handling properties. At the same time, the company undertakes a thorough patent search, which suggests that, although the formulation may be novel, the company must try certain other formulations as well, in order to cut off the possibility of later competition from an equally effective product. At the end of six months, the total cost of the effort has risen to $200,000.

Now the company files its patent applications and sets up a pilot production line. The line turns out to have a number of bugs in it; there are many more problems in producing large quantities of adhesive of reliable composition than in making small samples. Because of the scale of effort, modifications in the process became far more expensive than they had been before. Quality control becomes a problem. Shelf life has to be examined. Use tests, which had already been conducted on a small scale, are now undertaken on a large-scale, long-term basis. Meanwhile, word of the development has begun to spread outside the company and management decides to undertake a crash program in order to market the product before fall. A year and $600,000 have gone by.

By the time the first full production line has been set up, the marketing strategy has been settled, the sales force has been educated to the product, the use tests have been completed and evaluated, quality-control techniques have been established, two years have elapsed, and $1.2 million has been spent. Now the company can market the product — even though bugs in production, reliability, and quality still crop up. The product may or may not achieve the volume anticipated for it. The company may or may not make a profit.

The process described here takes no longer and is no more expensive than most product and process developments as they occur in medium-to-large firms. To be sure, there are instances of new products that have been invented, developed, and brought to the point of marketing for as little as $5,000. But at the other extreme are examples of products that took 10 years and $20 million. Regardless of scale, however, the shape of the development-cost curve remains relatively constant:

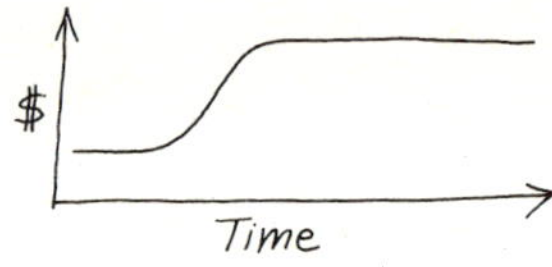

Often, the first invention — the first demonstration of an effect — takes no more than a month or two and a few thousand dollars. As the company takes its first exploratory steps, the rise in the rate of expenditure is slow. As the company passes further checkpoints, the rate of expenditure increases. The total commitment grows logarithmically. Each new major commitment requires complementary commitments. Only when the product or process reaches commercialization does the curve begin to level off.

The development-cost curve takes on meaning when it is put in the context of the corporate investment game and the conversion of uncertainty to risk. Despite the careful efforts of many companies to establish checkpoints beyond which they will not go without adequate justification, they find themselves having to make investment decisions on insufficient evidence in a general climate of uncertainty. As they begin to climb the slope of the curve, they rather quickly reach what appears to be a point of no return. At this point, the executive vice president can say to the president: "We have put in so much. We may as well put in a little more and find out whether the investment is worthwhile." To which the president can always reply: "Don't throw good money after bad." But when this point is passed there comes another point, where the mistake — if the development is a mistake — is too big to admit. Large-scale developments of the kind undertaken by supercorporations or the military may proceed for months or years beyond the point where they should have stopped; they continue because of massive commitments to errors too frightening to reveal. They have their own momentum. In these cases, the personal commitment of the people involved in the development, the apparent logic of investment, and the fear of admitting failure, all combine to keep the project in motion until it falls of its own weight.

The problem of innovation within the corporation is, therefore, a problem of decision in the face of continuing uncertainty. A man must take leaps — not once, at the beginning of the process, but many times throughout the process — always in the face of uncertainty and on the basis of inadequate information. The need for such leaps of decision grows out of the uncertainty inherent in the process. A company cannot escape it by careful planning, or by gathering exhaustive data. The uncertainties *resist* resolution — and the process of attempting to resolve them is itself a form of commitment.

Then why the rational view? Why the belief that innovation is manageable? One key to the answer lies in the very nonrationality and uncertainty of the process. A man may state the rational view when he is describing a process in which he has no part, or when he is trying to tell others how to do it, or to exhort others to do it, or, again, when he is reassuring himself about it. In short, the rational view may function as a device. It is an idealized, after-the-fact view of invention and innovation — as we would like them to be, so that they can be controlled, managed, justified. It is a view designed to calm fears, gain support, or give an illusion of wisdom.

The attraction of the rational view is easy to understand. Uncertainty is frightening. If the development of new products and processes is unavoidable — as growing numbers of corporations believe — it is cold comfort that most new products fail, that invention is full of unanticipated events, that innovation is inherently uncertain and subject to a treacherous cost curve. It is far more cheering, even apparently necessary, to believe that invention and innovation are rational, deliberate processes in which success is assured by intelligent effort. As a result, uncertainty becomes taboo, unmentionable — especially in the context of important corporate decisions. It is necessary to have a *clear, rational view* of where we have been and where we are going. It is necessary to believe that the future is essentially predictable and controllable if only we gather the right facts and draw the right inferences from them. We suppress the surprising, uncertain, fuzzy, treacherous aspects of invention and innovation in the interest of this therapeutic view of them as clear, rational, and orderly. Armed with this myth, managers make decisions and mobilize resources. They and their subordinates must then live out the actual uncertainty of this process.

But there is another answer to the question: Why the rational view? This answer focuses on its partial truth and on its utility.

The rational view of invention and innovation is more nearly correct for more nearly marginal inventions. The less significant the invention, the more the process tends to be orderly and predictable. The more radical the invention, the less rational and predictable.

I represent the universe of new products and processes with a target-like diagram:

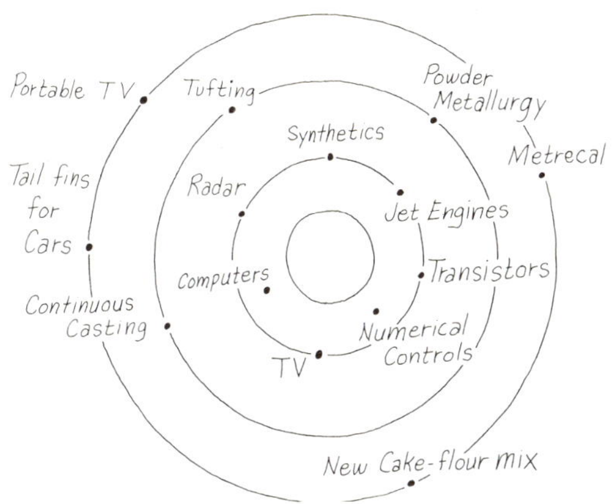

In the outer rings are those inventions whose acceptance requires least change. Indeed, these are hardly inventions at all. They require little or no change in the technology. They do not significantly advance the state of the art. They require no modification of scientific theory. Production equipment can be made to handle them with only minor change. Little or no rethinking of the corporate organization or approach is required.

As we move toward the center of the diagram, we encounter inventions that carry with them *major* changes in technology: transistors, synthetic fibers, etc. Their introduction goes hand in hand with change in scientific theory. They command new concepts in marketing and new marketing organization, as well as radically different production equipment and major new investment. They force their corporate organizations to undergo major change.

Let us take up the question of degree of novelty — how does this affect the organization within which innovation occurs? We can state a general rule here: The more peripheral the development, the less change required for acceptance, the more the development will tend to conform to the rational view. But the question of degree of novelty is more complex than this simple rule would suggest. For example, a technically insignificant invention — such as one that provides individually packaged food products — may require major changes in marketing. An the other hand, a major technological invention, as was true with the invention of nylon, may require surprisingly little change in textile machinery. There is an unusual optical effect: What looks like trivial change from afar may appear, close up, to be monumental.

For example, consider the replacement of a natural material by a synthetic — a disruption that may threaten several levels of a company at once: The established technology will be rendered obsolete. Men who have built up craft skills over decades will suddenly find their skills irrelevant to the new problems. Production will change from batch processing of materials to a continuous chemical process. New means of controlling product quality, production scheduling, and inventory will have to be devised. The far greater productive capacity of the new machines will require new marketing concepts. The old volumes will no longer be adequate. It may be necessary to double or triple the volume of sales, at lower prices, in order to keep the company's net profit constant. New accounting methods will be required. The very strategy of sales will have to change. In short, the business — and hence the nature of the corporation — will no longer be the same.

Not the least of the effects of technological innovation is on top management itself. For top management, the cumulative effects of all these changes may be overwhelming. If the president came up through the business and draws his confidence from his intimate knowledge of the details of the present operation, technological innovation threatens to throw him onto completely unfamiliar ground. He understood the old business, but he does not understand the new one. How can he manage if he does not understand the business he is in? Where is he to draw the resources of experience he needs in order to trust his own judgments? Is he to become completely dependent on the proponents of the new product or process? He is faced with a crisis of both self-confidence and trust in others.

In brief, technological innovation disrupts the stable state of corporate society. On every level, it affronts the society's vigorous and continuing efforts to stay as it is. To this onslaught, the corporate society responds in a variety of ways. Some of the more evident are these:

It rejects the effort at innovation. It puts down the idea, fires or demotes the men associated with it, and makes its further discussion taboo. Many corporate pasts are littered with discarded efforts of this sort — begun, then stopped, then relegated to corporate limbo.

It allows the effort to continue, but isolates it from the rest of the corporation. In this way, research projects — or whole research departments — may function in a vacuum, cut off from contacts with the corporation, contacts that are essential to their ever coming to reality.

It contains the threat, allowing it to proceed, but on a level so much reduced that the innovation is always far short of critical mass.

It seeks to convert the threat to an activity acceptable within the corporate society. Efforts at radical innovation become product improvements or service to production or sales, which can be carried on without disrupting effect.

These are straightforward negative responses. But the corporation cannot respond to innovation in an exclusively negative way. It cannot, because it believes itself to be committed to technological innovation, as essential to corporate growth. Innovation is something the corporate society must both espouse and resist.

It may compartmentalize innovation, permitting it to occur in one part of its business while preventing it from occurring in others.

It may oscillate between support and resistance, confusing corporate members by an on-again, off-again approach to change.

It may resist innovation while not appearing to do so, while in fact proclaiming the official doctrine of innovation. For example, it may encourage the development of new product ideas only to find consistently that none of them meets the

A new development shakes the boss's self-confidence

He may reject it

Or let it go forward, by inches

Or allow it to continue, in isolation

Or control it

Or give it lip service

stringent criteria laid down in advance (just as a mother may want her son to get married but never approves of the girls he brings home); or it may foster a research effort, only to reject or ignore its results.

These strategies do not usually represent a conscious pose or a management deception. They are responses of a corporate society to two requirements that are equally legitimate — even necessary — but unfortunately in conflict with one another.

The society of a corporation attempts to maintain a stable state. This effort is not inertia, but conservative dynamism. The various forms of corporate resistance to change that reflect themselves as obstacles to technological innovation are processes of *conservation* — processes that are essential to the survival of any social or biological organism. Active conservatism is the natural state of organisms. It is nonsense to say that companies should throw off this old-fashioned habit. But it is the crisis of the modern industrial corporation that it is required to undertake technological change — change that is destructive of its stable state — in order to survive. This is the paradox that makes the corporation so vulnerable. This is the paradox that accounts for its ambivalence to innovation and yet forces it to adopt new forms and styles of change.

Donald A. Schon

9
To Promote Invention

Is it true that the productivity of our large industrial laboratories is low, that the creativity of our scientists has fallen off? I have gathered a good deal of information on these questions, for as a consultant to technical management, and as one who has devoted many years to the task of managing research in industry, such information is of extreme importance to me. Specifically, I have asked the presidents of some two hundred of the nation's largest corporations to assess their firms' research activities: Were they satisfied with the productivity of their companies' research organizations? Most respondents expressed dissatisfaction, citing "lack of creativity at the bench" as the most significant cause of disappointing yields.

In this essay, I want to discuss the question of research productivity, particularly the productivity of the large industrial laboratories. If it is true, as so many company presidents suspect, that creativity is lacking, why is it lacking? And what can be done about it?

Let me begin my discussion by reciting an old line, which has been applied time and again to research — and which has done incalculable harm. It is the familiar quotation attributed to Ralph Waldo Emerson: "If a man can write a better book, preach a better sermon, or build a better mousetrap than his neighbor, though he builds his house in the woods, the world will make a beaten path to his door." There is some doubt as to whether Emerson ever said those words.

I hope he never did, for they have no relevance to today's research.

Two groups of people are victims of the better mousetrap misconception: those who invent mousetraps that they believe to be better, and those who employ the inventors. The inventor who believes that a new invention of value will "sell itself" is exposing himself to bewilderment, lack of confidence, frustration. When management does not eagerly grab his new development, he is not going to understand. To management, belief in the misconception makes it difficult to understand why its research organization does not come up with new developments for which there are obvious markets.

It is important for everyone connected with the research and development operation to realize that the actual products of the research department are only *potential* new products or processes — and nothing more. Before any of them can become commercially successful, they must first be sold to management.

Selling research developments requires skill, persistence, and courage. I emphasize courage because there is always danger of failure and the consequent damage to one's reputation. In most companies I have known, the man who is associated with a research failure is rarely given a "plus" from top management for his effort, so far as advancement is concerned. And often he is given a black mark. This is not the way it should be — but it is the way it is in most companies. To my knowledge, there has never been a successful new product put on the market that did not have its "product champion" — someone who risked his reputation, or possibly even his job, to put it over.

Management that wants to succeed in increasing the productivity of its research operation must always be aware of the essential role played by the product champion. And it must see to it that the barriers he encounters are not insurmountable. There must be barriers, of course, for they constitute the screening operation that separates good projects from poor ones. But in many companies these barriers are so formidable that good developments are lost — due either to

AFFIRMATIVELY NO
AMEN
NOW WAIT A MINUTE
DONT EXPECT WE WILL
NO!
WOULD YOU BELIEVE NO ?
GRIVLE ...
SORRY I'M NOT IN!
OK?
YEAH!
HOW LONG HAVE YOU BEEN WITH US?
NYET
HAVE YOU WRITEN... HOME?
I Dont Care What your Mother Thinks
STOP
I LIKE YOUR APPROACH BUT
WHY DONT YOU DO SOMETHING NICE?
Send a memo
No
NEG. AND THATS POS.
NO NO
NO
DO YOU ... HAVE AN ... UH ... AN APPOINTMENT?
CHUCK IT
SORRY 'BOUT THAT
I DON'T THINK SO
NO!
NO!
NO
STOP
CORPORATE DECISION MAKERS
SEARCH
CORPORATE RESEARCH LAB
DESCHAMPS

lack of courage or incentive, or both, on the part of those who must sell those developments to management.

Some projects are easier to sell than others — an obvious statement. But if management fails to take this fact into account — which often happens when management is not experienced with research — its research program becomes lopsided, favoring easy-to-sell ideas, as Donald Schon pointed out in his discussion of innovation. This accounts for the nonproductivity of research in many large companies. How do we account for this? Partly, it stems from the well-implanted mousetrap idea. But more than this, it is due also to another well-implanted idea (also untrue), which is expressed by the old saying: "Necessity is the mother of invention." These two cliches add up to a conviction, perhaps unconscious, that if a new invention has any value it will have obvious utility in filling an existing need. And here we come back to the question of lopsided research and why it occurs. Inventions that fill obvious needs are easy to sell. Likewise, projects that have as their objective the filling of an existing, well-recognized need are also easy to sell. Thus, when management has a reputation, among its inventive people, as a timid soul in terms of its receptivity to new ideas, only the easy-to-sell ideas will come forth. In such an organization, a selective process is going on: The research program becomes heavily laden with easy-to-sell projects, projects that are oriented toward filling obvious needs.

Many research managers *insist* on "need-oriented" research as the way to insure productivity. But in fact, is this type of "practical" research really the most rewarding in terms of improved profit margins? Products like neoprene, nylon, polyethylene, silicones, penicillin, Teflon, xerography, and the Polaroid Land Camera did not come into being because the public demanded them. That happened only after the product was developed.

Most of the products I have just listed proved to be exceptionally good profit earners for their companies. They are the kinds of products most managements really hope for from their research programs. But they are unlikely to come if the research effort is limited to filling existing needs.

How do you get such new products? I believe you do it through "practical" research, but what I mean by the word *practical* may be different from that of most research managers: To me, practical research can be one of two types. The first is research directed toward filling "envisioned" needs — as distinguished from obvious needs. For example, Edwin Land envisioned a need for a camera that would give the user a picture in one minute. He believed that if he could invent such a camera, he could develop a market for it. He was surely right. The other type of practical research, which gave us many of the products just mentioned, is carried out for the purpose of exploring new fields of science. It is not research for the sake of research; you do have some long-range objective before you. But this is "discovery" research, for it is analogous to geographic explorations. To be sure, many of the discoveries may turn out to have only academic significance, but such research is necessary if you are to develop such things as nylon, polyethylene, transistors, or penicillin. My point is simply that to ensure maximum productivity from industrial research, three types of research should be going on in the laboratories: There must be research whose aim is to fill envisioned needs. There must be research aimed at making new discoveries. And among with these there will be the customary research that is oriented toward filling existing needs.

Now let me ask you a question. If Edwin Land had been working in your laboratory, would he have got the money for his project? If Chester Carlson had been working in your laboratory, would he have been allowed to perfect his Xerox process? By hindsight, your answer is yes, of course. But let us take the example of Carlson. He took his Xerox invention to well over 20 companies before he was able to make a sale. No one except Carlson himself envisioned the need for his process or believed it was practical. Since more than 20 companies said "no" to Carlson, this means that if he had been an employee inventor, with but one chance to sell his idea, his chances would have been less than 1 in 20. If we are looking for increased productivity from research, a 20 to 1 chance of success for a novel, profit-making invention is just too low.

The problem seems to be in selling the invention after it has been made. How do we develop the determination and persistence that are necessary on the part of the product champion to achieve success with our research?

We must understand the problem. We must try not to make the barriers insurmountable. We must try to promote the selling of inventions and minimize the hazard of the black mark if the inventor fails. But we must also provide an *incentive.* Incentives are essential to progress in every field, and research is no exception. Let us take for illustration a large industrial research organization, where a product champion is attempting to sell a research development to management. What is management? Usually, it is several levels within the research organization itself, plus others in the marketing area, plus the top levels of management, where the ultimate decision will be made. In such an organization the inventor may find that many of the people to whom he must sell his project are slow to arrive at a decision; he will find too that those decisions are usually negative ones. In most organizations, negative decisions are less risky. Indeed, there may be no incentive to behave otherwise. Promotion will come, more often than not, to the person who has not been associated with a failure. Most surely it will come to this person rather than to the man who has tried but failed. To overcome this natural bureaucratic tendency, powerful incentives are needed. And some successful companies have devised such incentive plans. In particular, the plan of the Du Pont company is especially effective. Why? Because it receives the full attention and support of top management. Indeed, as a practical matter, top management must be the originator of such incentives. Unless this is so, and unless top management *works at* keeping the incentive system functioning, there can be no effective means of encouraging an increase in research productivity. In general, such incentives will comprise a clearly defined policy of promotion based on evidence of courageous decisions that lead to successful new products and processes. But an incentive plan must do more than this. It must also recognize and reward the courageous individual whose idea led to an *unsuccessful* commercial product. Call him a failure if you wish, but this man is more valuable to the organization than one who continually makes safe, negative decisions. William L. McKnight knew this and he used it as he built the 3M Company during its greatest years of growth.

In attempting to improve the innovative process, management at all levels must also know that all new developments have weaknesses. It is easy to cite what is wrong with a new development — almost anybody can do this. But what is harder to do, and more valuable, as psychologist Ray Hyman showed us in his essay on problem-solving, is to recognize the good qualities of a developing product. When you have recognized these good qualities, you then go on to try to overcome the product's deficiencies. All too often we reward the manager who is so skillful at the easy part of this process, the man who knows how to find deficiencies and thus saves us from making "serious and costly mistakes." This is the sort of fellow who rises to the top in many corporations.

I should like now to turn to the question of the corporate incentive, and how it can be improved. How can the corporation be encouraged to support more research — and riskier research? First, we must recognize our targets. We want to accomplish two things:

We want a maximum stimulation of invention to insure technological progress.

We want maximum utilization of the worthwhile products or processes resulting from this technological effort.

These were always the goals of any patent system, even in ancient Venetian times, when the first such system was devised. But we must admit that some things have changed in the intervening 500 years. Indeed, there have been significant changes in these past 50 years, as the United States has passed from an agrarian age to a highly industrialized one. In this change, we find that the number of significant inventions made by independent inventors has dropped markedly. The independent inventor has given way to the paid-to-invent employee working in an organized research laboratory, a result of the complexities and costs of modern science; the independent inventor cannot afford the tools. Whether in an independent company or in a laboratory of the government, these research people have had to sign away their rights to any patents that issue. This is a condition of employment. And this is as it should be, for otherwise there could be no jobs for those inventors.

But this change does affect the incentive provided by the patent laws — laws that were designed to give patent rights to the inventor. Now, with the rights going to the employer instead, the original incentive is lost: No longer does the inventor reap the financial reward, unless — as in a few firms — his company itself rewards him. But there is an even more serious loss, and that is the ability of the inventor to sell his invention. As I said earlier, he has but one chance: He must sell the idea to his employer. Beyond that, no matter how much faith he may have in the value of his invention, he is powerless to do anything with it, for it does not belong to him. Not only may this result in reduced incentive to make more inventions, if previous ones have not been utilized, but it may also result in another failure, inherent in the original concept of the patent law: The public does not get the benefit from the invention for which it authorized the monopoly. It was not the original intention that the granted monopoly carry with it the right *not* to make the new invention available to the public. Rather, the idea of granting the monopoly was precisely to serve as an incentive to commercialize. The incentive to invent, of itself, was only of secondary importance, for in order to get a monopoly position it was necessary first to invent something new and useful. This was the important thing. And in my opinion, this is the area in most need of change as inventors have changed from independent to dependent status.

With no one to sell to except his employer, the inventor nowadays can only suggest that his idea is worth developing. In most cases, it is someone other than the inventor who decides its worth to the company. This someone, being more "objective," may not see the potential that may actually be present. It is my conviction that a good many worthwhile inventions have been lost to the public in this way. What is lacking is the incentive for the company — the owner of the patent — to do something with the invention it owns. I do not believe, as a general rule, that companies deliberately "sit on" patents or find it advantageous not to market a new invention because it may detract from a product they are presently manufacturing. Unquestionably, there are cases where this has occurred, but a much more frequent cause of failure to commercialize a patented invention is simply lack of vision — failure of those empowered to make decisions to see the potential of the invention.

What is the present situation regarding incentive for the employer to commercialize the patents he owns? In the first place, he owns those patents as the result of an agreement with his paid-to-invent employees: He furnishes the tools and he pays their salaries. In return, they assign their patent rights to him. And this is as it must be. Ownership of the patents by the employer is absolutely essential in our present-day world.

If the system is changed, we must not only preserve the ownership of the patent in the hands of the employer, but we must somehow provide for an additional incentive — equal to that which the independent inventor has. This might be done by providing that, if the corporate owners of patents do not commercialize those patents after a period of five years, they must forfeit their ownership on demand.

This is not a new idea. It is law in most European countries. The intent of the law is to correct precisely the situation I have been describing. The European laws take the form of compulsory licensing of patents, or the payment of maintenance fees to keep those patents in force. Presumably, the owner will only pay these fees if he is using — or intends to use — the patents.

There is only one difficulty with this solution, however. It does not work. And the reason it does not work goes back to the mousetrap philosophy. The world is *not* eagerly waiting for the potentially valuable but unproven invention. Carlson had to *sell* his Xerox invention — and those 20-plus companies were not eagerly waiting to be sold either.

The trouble with this scheme — and similar ones that are frequently proposed — is that it fails to take into account the fact that a better mousetrap is not immediately recognized as such. It is not sought out by entrepreneurs. And such schemes have another, and even more serious, trouble inherent within them: Any scheme that takes away ownership of a patent from the original corporate owner and gives it to the public — "fair" as this may seem — *insures* that new inventions will *never* be made available to the public. Why not? Because all incentive to support invention will now have been destroyed. Who will put up the capital, perhaps millions of dollars, to develop an idea and establish its marketability, when after he has done this anyone may copy it at little expense? The monopoly has been destroyed and now the inventor, dedicated and enthusiastic as he may be, has nothing whatever to sell.

What must be done is to increase the incentive, not destroy it. The way to do this is to take the patent away from the owner who has not commercialized it, but at the same time preserve the monopoly. This means compulsory *exclusive* licensing. That is to say, after a five-year period, if the corporate owner has not commercialized the patent, the inventor may attempt to find an entrepreneur who is willing to invest in the commercialization of the invention. Now the inventor has something to sell, for the entrepreneur can request an exclusive license, which is now mandatory, and ownership of the patent can in effect be transferred to the entrepreneur. Thus the entrepreneur can invest with assurance that he has a reasonable period in which to get his investment back with a profit if the venture is successful. To be fair to the original owners of the patent, who financed the research leading to the invention, the compulsory license should also carry with it the right of the original owners to receive a reasonable royalty. Now we have been fair to all parties, have not confiscated private property, and have provided for alternative routes by which an employee inventor may bring about and utilize his patent. This very nearly puts our employee inventor in the same position as the independent inventor. A maximum of incentive has been introduced. Inventors who are frustrated as the result of seeing their previous inventions gather dust can now become reactivated. This might well bring back into circulation many creative individuals whose ingenuity is no longer utilized. But of equal importance, we would have greatly increased the incentive for the original owners to commercialize the invention.

As things stand now, if a corporation is sitting on a Xerox-type patent, for whatever reason, it has little to worry about. If that patent were to revert to public ownership for failure to commercialize, its corporate owners would still have nothing to fear, for it is certain that invention would not be developed. But under a compulsory exclusive licensing plan, there is a very good chance that some entrepreneur would avail himself of the rights to that patent — and perhaps develop another Xerox. The fear of being "shown up" in this way will serve as an effective stimulus to reconsider any and all patents approaching the five-year point. I can think of many problems that might result with such a plan — just as the negative thinker can with any new idea. For instance, what do you do in a situation where a company has developed several commercial patented processes for the manufacture of a new product? For obvious reasons, the company commercializes only one of these. It would be unfair, indeed it would defeat the purpose of compulsory exclusive licensing if the company were forced to license its alternative patented processes because it had not commercialized them. A solution to this problem would be to provide that, if the company commercializes any one of a group of related patents, it would be entitled to keep the whole group. The point is that this is a new idea — and, just as with a new product, we have to recognize its advantages and devise ways to overcome its deficiencies.

Carl E. Barnes

10

Loyalties

If his organization is unwilling to sponsor work that will lead to the type of professional recognition the research man seeks, he must change the organization, modify his own aspirations, or go to work somewhere else. No matter which course of action he chooses, his research manager is caught in the middle of the problem.

The sense of professional identification possessed by scientists, and their desire to achieve recognition and reward within their profession are well-known qualities. But the man who assumes his first position as a manager of a research group sometimes underestimates the difficulty he will encounter as the man in the middle. Typically, his preconceptions of his new position will emphasize the "technical" aspects of his job. He will assume, with some justification, that his selection was heavily influenced by his demonstrated competence in conducting research, and his technical judgment and experience. Quite naturally, he will expect to be deeply involved in determining the technical program of his group, in guiding the conduct of the research, and in evaluating the success of the effort. He knows he will spend considerable time in "communicating and coordinating"—in maintaining effective relationships with corporate management and in coordinating the efforts of his group with other technical groups. If he is realistic, he will also expect to devote effort to insuring that the output of his group is properly appreciated and that it moves successfully along the road toward eventual production. Finally, the new manager of a research group will probably be resigned to doing the routine paper-work that most organizations seem to believe must be done.

All of these are activities that he likely has had some experience with. And, to be sure, they are indispensable parts of his job. But the real test of a research manager is made on the basis of how well he performs in *both* jobs — the "indispensable" responsibilities as well as those of the man in the middle. The successful manager of research spends a significant amount of time and effort in this second job, where his previous training and experience in research are of limited value. What he must learn to do is establish and maintain a viable accommodation between the needs and aspirations of his research people on the one hand, and the demands of his sponsoring organization on the other. In this uncomfortable position, the manager is constantly subjected to conflicting, sometimes incompatible pressures. In some fashion or another, he must mediate them.

The research manager is not unique in this requirement. The establishment of areas of mutual interest between the goals of the individual and his organization is now regarded as a major part of any manager's job. But the research manager does face the problem in more acute form than do managers in other functions.

The people he deals with have substantial mobility and a highly developed sense of personal autonomy. Identification with their profession takes precedence over identification with their employer.

The greater the divergence between the goals of the individual and the goals of the organization, the more acute the problem of the manager becomes. If attractive alternative sources of employment are available to the research man, his manager's problems are further compounded.

The difficulty of establishing areas of mutual interest between the individual and his organization depends on the extent to which their two sets of interest converge. This problem can be looked at in two parts: One is concerned with the explicit technical outputs the organization expects when it hires a man, as compared with the man's personal goals for professional accomplishment; the other, a more subtle area of conflict, involves the needs, aspirations, and value system of the individual, as related to the cultural and social system of his organization. Every organization develops such a system, and out of the system are generated the organizational requirements concerning acceptable attitudes, beliefs, and ways of behaving.

The technical outputs the organization seeks when it hires a man may cover the spectrum from short-term accomplishments that will reduce the cost or improve the performance of current products to major scientific advances that may open major new business opportunities. But in general, as both Schon and Barnes have said in their essays, the organization will be more receptive to shorter run, rather than longer run achievements, and lower risk rather than more speculative programs.

Converting these generalizations into a specific program that represents an optimum expenditure of funds requires great judgment and ingenuity. The problem would be difficult enough if the manager were omniscient, or if he were dealing with robots. But instead, he is dealing with individuals who have a highly developed sense of personal autonomy, whose personal and professional goals are not easily reconciled with the organization's expectations.

Unfortunately, the overlap between work that is professionally significant and work that has sufficient commercial significance to warrant sponsorship by the organization is by

no means complete. The scientific fraternity does not apply economic criteria in judging the value of scientific accomplishments. Thus, the manager is continually effecting delicate compromises. He is striving with all the wits at his command to ascertain the possible commercial value of work that is professionally significant and, at the same time, trying to formulate programs of commercial value in such a way that they may have some professional significance. In addition, he is trying to juggle time so that people will, over a period of time, have the opportunity to do work of both types if they cannot do both simultaneously.

While the manager is effecting these compromises, he is also exerting pressure on both the individual and the organization. He is trying to modify the aspirations of technical people so that they perceive acceptable avenues for personal achievement by doing work the organization is willing to sponsor. He is also trying to educate the organization to the commercial value and the necessity, from the standpoint of retaining highly competent people, of longer range work that is more professionally satisfying.

In attempting to modify the goals of both the professional people in his group and the sponsoring organization, the research manager exposes himself to considerable risk. In effect, he must assure both groups that a modification in their respective goals will be worthwhile. In the case of the professional people, he is asking them to adopt new avenues for satisfying one of their basic human needs: esteem, both of self

and from others. This esteem is expressed, on the one hand, in feelings of strength, adequacy, competence, etc.; and, on the other hand, in terms of prestige, reputation, influence, and attention.

It is apparent that these latter feelings exist in a social context; they involve comparisons with some reference group. To the extent that the manager asks people in his group to perform work that would not elicit esteem from the scientific fraternity, he must assure them that adequate recognition will be forthcoming from another reference group, that is, the sponsoring organization. He must insure that the sponsor will provide adequate recognition, reward, and influence to compensate for the equivalent satisfactions that the individual would otherwise seek from the scientific fraternity. Since the manager himself has only limited influence over the sponsoring organization, he is in danger of promising his people too much. If the scientist is to be induced to seek esteem, at least to some extent, from another reference group, he must be convinced that equivalent satisfactions are attainable. If he is misled, disillusionment is certain to follow.

To the sponsoring organization, the research manager makes assurances that its support of longer range, more professionally significant work will be a worthwhile investment. In view of the increased length of time before payoff, and the inherently speculative nature of such work, it is clear that the manager is at risk in promoting programs of this type. He must have a management that has some appreciation of the nature of his problems and that has worked through the difficult problem of formulating its basic posture regarding the role of research in the enterprise.

Quite naturally, continual exposure to these conflicting pressures places a great emotional burden on the research manager. He faces a personal handicap in dealing with them, because he typically will feel ambivalent about the very issues he is trying to resolve. The manager of research and development almost always has a technical background and was once an active contributor to science or technology. Thus, at one time he had aspirations and values very similar to those of his people. The move into management usually involves a reorientation in aspiration and values — a reorientation that generates discomfort. In resolving this personal problem, the manager is in danger of losing his sensitivity to the needs and aspirations of his people. Furthermore, he has other, still more subtle, areas in which to establish an accommodation between the individual and the organization. These involve the personality structure and the value system of the individual and the culture and social system of the organization.

Many of the problems in the search for accommodation come into focus in two areas: the acceptable limits of organizational loyalty and the meaning of authority and responsibility. Most members of corporate management place a premium on organizational loyalty. Their own commitment to the goals of the organization is very heavy indeed. Furthermore, they identify their personal fortunes and their sense of personal accomplishment with the success of the organization and their ability to contribute to that success. Quite understandably, they are uncomfortable in the presence of employees who may question the goals of the organization, or the efficacy with which they are being pursued. Employees who diligently seek to do work that will lead to status and influence in groups outside the sponsoring organization also become suspect.

On the other hand, scientists and engineers tend to chafe at what they regard as an unwarranted requirement to identify with the organization. Their fealty does not extend to uncritical acceptance of managerial policies, decisions, and actions. They view with alarm the possibility of becoming so dependent on their employer that they lose their independence. Many strive to preserve their marketability by doing work and carrying on professional activities that will gain them external recognition.

Much of this difficulty arises from corporate management's lack of appreciation of the nature of the scientific mind and of the characteristics of discourse in science. There is no reason to believe scientists will drop the habit of critical questioning and skepticism at the door of the laboratory, or that the goals of the corporation and the performance of management will be exempt from such critical questioning. What is not so generally recognized is that the discourse of scientists is frequently harsh and intense. They express their feelings, beliefs, and disagreements bluntly and in unmistakable terms. To one who is not familiar with this phenomenon, the opinions and feelings expressed may seem disloyal.

The phenomenon that is frequently interpreted as disloyalty or lack of interest is the scientist's highly developed sense of personal autonomy. The scientist is more inclined to ask the organization to accept him on his terms. As social scientist Anne Roe has pointed out, the scientist has an intense desire to control his own work and derives great satisfaction from being able to do so. This personal need creates another area of conflict, which the research manager must seek to resolve.

The current uneasy status of the concept of authority among organization theorists comes into sharp focus in the role of the research manager. This can be stated simply as the conflict between authority based on knowledge versus authority based on hierarchical position. The whole tradition of science stresses the indispensable requirement that authority in science be based on knowledge. Without it, science can be perverted and stultified by political, religious, and social pressures.

On the other hand, in spite of advances in sophistication, organization structure is still basically founded on the authority of position. Historically, this authority of position was buttressed by — and in fact legitimized by — the authority of knowledge. Increasing specialization has tended to create a gap between the two. This is particularly true in research where the manager is dealing more and more with specialists who have far more detailed and sophisticated knowledge in their particular fields than he does. The question of what now legitimizes the authority of position is becoming increasingly urgent. It is deeply galling for a scientist to defer to the manager's judgment, not because he is right but because he is boss. The naked exercise of power clearly is not the answer, for, as we will see, the scientist has too many attractive alternatives.

Clarification of the nature of the responsibility and authority of the manager, as well as of the people in his group, must be achieved if the legitimacy of hierarchical authority is to be re-established. The perceptive scientist recognizes the dilemma his manager is currently in. So long as the manager is held responsible for the output of the group, he must ultimately have control over it.

The concepts of responsibility and authority have historically been treated almost as physical entities that can be subdivided into pieces. Under this old concept, if one individual, within a given component, has more authority and responsibility, another individual must have less. The emphasis is placed on the most effective division of the authority and responsibility pie, not with the size of the basic pie to begin with.

But there is another way of viewing responsibility and authority, as you will see in Rensis Likert's essay at the conclusion of this section. These are not quantities to be subdivided. Neither are they additive psychological entities. An individual may perceive that he has increased responsibility and authority without believing that his manager now has less. In fact, he may feel that his manager has more, too. It is coming to be recognized that people's perceptions of these relationships are what count, not the arrangements specified in an organizational hierarchy. The skillful research manager spends considerable time and effort establishing a concensus with his people on the legitimate exercise of authority. He recognizes the deep need of his people to exercise control over their lives and work. At the same time, he relies on their sense of integrity. Scientists do not wish to be employed as noncontributing showpieces, nor do they deliberately act irresponsibly. They believe they can make important contributions to the formulation of programs and the general effectiveness of the organization if given a chance. Making use of their talents contributes enormously to the strength of the group, but it calls for a high order of skill, perception, and emotional maturity on the part of the manager.

Decisions are rarely dramatic acts completed in lonely isolation. More typically, they emerge from a complex influence process to which both the manager and his group can make vital contributions. If the manager has the courage to let his people really become a part of this process, they will become less concerned with who is deciding what and more concerned with the soundness of the decision. The technical people will become more committed, more deeply involved in the success of the total effort. They will also be exposed to experiences that will contribute to their individual growth and thus to the collective competence of the group.

By seeking to exercise less personal authority, the manager increases the effectiveness of the group and, therefore, enlarges the total authority and responsibility exercised by the group. Effective groups are listened to, they are supported in taking on more challenging work, and they receive more funds.

Nobody should infer from this discussion that if the research manager performs this role successfully he will find himself in utopia. In fact, he frequently may feel he has less freedom and authority than the members of his group. He constantly must balance the possible cost of accepting a decision he believes is unsound against the reduction in mutual respect and individual involvement that the arbitrary exercise of authority will create.

This modified view of the nature of the manager's role is consistent with changes that are emerging from recent studies of managerial effectiveness, including those of Pelz and Andrews, which you read about earlier in this volume. Likert tells us later on that increasingly the manager will be responsible for establishing the environment in which people work and that responsibility for technical decisions will rest with specialists. The pressures inherent in their role are forcing research managers to pioneer in evolving new patterns of behavior.

The pressures are further reinforced by the current situation regarding the employment of scientists and engineers. For the last decade or more, the demand for them has been consistently high and this situation is expected to continue. Furthermore, these employment opportunities cover a wide range of careers that can appeal to a broad spectrum of individual goals and aspirations — universities and colleges, industrial research and development, government laboratories, nonprofit laboratories — all are offering employment.

To compound the problem, these people are, as a class, unusually mobile. In most cases, ties to parents and the local community already have been broken. In addition, the cost of relocating is usually borne by the new employer. Consequently, scientists and engineers face neither serious emotional nor financial hurdles in contemplating a change in employment.

Finally, these people are remarkably well informed about employment opportunities. Knowledge of prevailing salary scales is widespread and easily attainable. If he wishes, the individual can easily and unobtrusively get a reading on his current market value by a few discreet inquiries among professional friends or by keeping his ears open at a professional society meeting. The grapevine operates with remarkable speed in broadcasting the changing fortunes of various research and development organizations throughout the country.

All of these circumstances combine to magnify the burden of the research manager in establishing an accommodation between the needs of the individual and the goals of the organization. If he is not successful, he is in danger of losing the man altogether, rather than merely having him fail to contribute his fullest. In the next essay, I shall take up the question of what the individual expects of his boss in this sensitive relationship.

I began by saying that a research manager has two jobs: one technical and the other involving the establishment of a viable accommodation between the organization and those in his group. It would perhaps be more accurate to say he has two alternative ways of approaching his job. One, which at least initially is more satisfying to his own ego, emphasizes his personal contribution to the formulation of programs and the conduct of research. The other stresses enlisting the total resources of the group to the establishment of goals and the accomplishment of objectives.

This latter approach contributes enormously to the growth of individuals and, therefore, in the long run, increases the resources available to the group. It frequently is argued that scientists are too specialized, too parochial in outlook to be trusted with such participation. Maybe some of them are, but all too frequently they do not have the opportunity to learn, to develop the point of view that makes it possible to contribute. Responsibility is a very sobering thing. People with intelligence, integrity, and drive will not act irresponsibly once they are, in truth, participants in the ongoing life of the organization. They will make mistakes, they will have to learn, but given the chance they can increase enormously the capability of the organization.

Lowell W. Steele

What's the Boss For? II

Scientists and engineers have considerable influence on how they will be managed — probably more than most people. They know that the demand for their services is strong. And they know that this factor, combined with their own personal mobility, means that management cannot ignore the possibility of losing their services. More important, the very nature of technical work — creating ideas, manipulating esoteric equipment and analytical techniques — gives the technical man great freedom to determine what he will do, how effectively he will do it, and how hard he will work.

When his talents are fully mobilized, he is capable of remarkable dedication and involvement. But these same talents, if negatively motivated, can be immobilized by indifference, and even diverted to subversion and subterfuge.

Much of the success of the manager of technical people depends on his skill in interacting with the members of his group. He must be able to interpret correctly the information available regarding their feelings and reactions — information that is usually incomplete and sometimes even misleading.

One important insight he must acquire is this: How do the people in his group view his role? What functions is he perceived as performing? What importance is attached to these various functions? Most studies of technical people pay little attention to this aspect of technical management.

What I shall do in this essay, then, is talk about the boss as he is seen by the people who work for him. My observations are drawn from conversations with hundreds of scientists and engineers in research and development activities within my own organization — the General Electric Company. These interviews go back over several years and involve people in many locations, in many technical fields, in both commercial and defense-oriented work, and in both applied research and development. The purpose of these interviews was *not* to find an answer to this question of how people view their bosses. The purpose was to gain a broader understanding of many questions relating to research and development activities within the company. The emergence of some uniformities in the ways in which these people view their bosses was an unexpected by-product of those interviews.

I made no attempt to elicit responses to a structured list of questions, but rather, in the course of an extended conversation, I would ask, "What is your boss's job?" . . . "What is he there for, as far as you are concerned?" . . . "What do you talk to him about?" I would simply give a person an opportunity to make any observations that might occur to him. I see this procedure as a sort of "through the looking glass backward" view — a view of the technical world from the inside out.

The responses I thought most interesting were those of mature, competent people — the kind who constitute the backbone of any organization. I will draw more heavily on these than on comments of new people or people of only marginal competence. The latter will only be used occasionally, for contrasting views.

The responses are interesting for the things omitted as well as the things emphasized. For instance, virtually nobody talked about his boss's relationships with *other* members within the immediate group. And little was said about allocation of resources within the group, or about organizing the work of the group, or about influencing interpersonal relations, or about facilitating communications. These topics are frequently discussed in more formal management seminars and the like. Why their omission here? Their omission may reflect the fact that such activities have "low visibility" from the viewpoint of the working scientist or engineer. Or it may indicate that these scientists and engineers attach little significance to such activities, or that they prefer to depreciate their significance.

Perhaps the most striking omission in people's responses was the absence of references to the manager's role in formulating the program, i.e., the technical work on which the resources of the group are focused. Traditionally, this function is regarded as *central* to the work of the manager — both in the broad policy sense of determining the basic objectives of the group and in the more immediate sense of deciding what programs will be pursued and at what level of effort. But the group members seem to see things differently. Mature, competent people do not see the manager as playing an important part in program formulation, or else they prefer not to ascribe to him any significant influence. It is likely that it is a subject on which they feel some ambivalence.

An interesting contrast is derived from the comments of people who were new to the organization or whose competence was marginal. Although relatively few of these people were interviewed, I found that they were more inclined to mention the importance of the manager's role in determining the program and in inspiring and leading people. Their boss loomed as a larger figure in their lives.

The possibility that mature technical people feel some ambivalence on this subject, and thus may tend to repress consideration of the impact of the manager on the formulation of the program, is reinforced when one examines the responses of really senior people. Most technical organizations have one or a few key technical people whose contributions are regarded as being in a class by themselves. These people have high status and they know it. They also have competence and reputation that are readily marketable, and they know that too. Consequently, they have less need to be greatly concerned about their relationship with their boss. In fact, in some cases it is the *boss* who is concerned. Such people exhibit little tendency to suppress feelings regarding their relationship with their boss and tend to view their boss with a certain detachment. Some feel their boss doesn't know what to do with them. One man said, "If I were to tell my boss that I was leaving, he would weep and wail and say how terrible it was, but down inside he would be relieved, because he doesn't know what to do with me." Others indicate opposition to managerial involvement in their technical work. As one man put it, "When my boss starts poking his nose into my technical work, then is when we cross swords."

The fact that technical people do not ascribe an important role to their boss in the formulation of the technical program or in providing technical leadership does not mean that he is excluded from consideration in these activities. Quite the contrary. He is given a very important role, but one that is very different from that usually thought of. For want of a better term, we might describe his role — as his people see it — as one of "technical followership." Technical people, apparently, do not perceive their boss as making significant contributions in the way of conceiving ideas, suggesting experiments, criticizing results, interpreting data, etc. Or it may be that they find it more agreeable to ignore or depreciate his role. In any case, they regard these areas as *their* responsibility. They sometimes take care to point out that when they discuss these matters with their boss, they talk to him as another technical man with specialized knowledge in given areas, but not as the boss. They believe the initiative should rest with them in generating their own progress or in discovering sources of aid among their peers.

What scientists and engineers *do* want from their boss is the ability to follow the work as it progresses, understand its subtle nuances, appreciate its significance, and perceive its relevance to the objectives of the organization. They regard this activity, by their managers, as crucial to the process of evaluation that does and must go on. The man in research or development knows full well that he and his work must be judged, that decisions must be made regarding his worth to the organization. He knows that his boss must evaluate the promise of his work in comparison with the work of others or in comparison with other programs on which he might work. But how does he see his manager's evaluative role? He views it in highly personal terms. He views it as an *ethical* responsibility of his manager to *protect* him, not as an institutional responsibility where his manager acts as an agent of the *organization.* His manager is *his* agent, *his* representative. He regards his boss as having an ethical obligation to be familiar with his work and to understand its significance in order to protect his interests when his work is being evaluated and decisions are being made that affect his future.

In an industrial laboratory, evaluation of a man is focused in the hands of a few people. The man himself, though he is the individual most concerned and with the most at stake, is not directly involved in this process. Hence he recognizes that his boss's judgment of his work is likely to be the dominant factor, both in determining how he is regarded by the organization and in influencing his future with it. Consequently, he is vitally interested in his boss's competence to render knowledgeable, balanced judgments. He finds it rather unsettling to have to lead his boss step by step through his work, to have him ask questions that indicate that he doesn't really understand what is going on. What is even worse is to have the boss get enthusiastic about what the man regards as the wrong things.

Technical people do recognize that the evaluative aspect of the boss's job is a very demanding task in itself. They recognize the diversity and volume of activities being conducted under their boss's realm of responsibility and the tremendous volume of technical work going on in the world at large. They know his task of maintaining up-to-date "state-of-the-art" knowledge is virtually a full-time job: He must read in a variety of fields; he must cultivate acquaintance with many people scattered over the world, and he must be aware of new techniques and their applicability, and so on. Technical people appreciate the enormity of this job. BUT . . . they *expect* the boss to have all of this *and more* at hand, or available by consultation, when he judges their work.

How do technical people cope with what they regard as incompetent managerial evaluations? At the limit, of course, they quit. But more subtle effects can also have a deleterious influence on the productivity of the group. The most important is on the program itself. People can come to feel that the organization isn't equipped to protect their interests, so it's up to them to look out for themselves. One alternative is to do work that increases their marketability, even though it is not necessarily the most important work that could be done from the point of view of the organization. In this way, the individual achieves greater control over his own welfare, and he may even increase his stature in the organization, in spite of his boss. While concern in this area may be quite acute in individual cases, it did not appear to be wide-spread. Few people expressed concern that the boss was not protecting them. Many indicated sympathy for the magnitude of his task. Frequently, they told me they tried to bother the boss as seldom as possible, because they recognized the demands on his time and wanted to be as little burden as possible.

Technical people recognize that their work, and in fact the work of the entire group, is being sponsored by some sort of client or customer. It may be a government agency, an operating component of the corporation, or the top executives of the company. Somebody has to be sold on the merits of the work when it is being proposed, has to be kept informed and interested in the work as it progresses, has to be led to appreciate the potential of accomplishments realized, and has to be induced to invest additional resources in further development when breakthroughs occur. To some extent, of course, the individual scientist may be able to participate in this process. However, there are many circumstances in which the manager must represent the individual or the entire group. He must describe the work, interpret its potential significance to the company, evaluate the probabilities of success or failure, and defend it against criticism or attack. To the extent that he is successful, the individual or group will continue to receive the resources needed to carry on. And here again, the individual finds his fortunes resting in the hands of his boss. One man summarized the comments of many people regarding this managerial role with the succinct comment, "He's my agent" (the meaning is similar to the performing arts).

What does our technical man expect of his "agent"? He expects technical competence, facility in communication, and enthusiasm. He expects his boss to have technical competence so he will perceive the potential of the work and be able to appraise the probabilities of its success. Perhaps even more important, he sees technical competence as providing his boss with a sound base for the self-confidence needed to present and defend our man's proposal to a critical audience. And, while it may be a rationalization, technical people sometimes suspect that their bosses' lack of enthusiasm for a particular project reflects the bosses' own uncertainty in the technical subjects involved and an unwillingness to expose this weakness to higher echelons in the organization.

Skill in communication is one that technical people feel ambivalent about. The folklore of science tells them that the work should "sell itself," that it is somehow improper for a scientist to turn himself into a salesman, even though examples that refute the folklore are plentiful and well known, as Carl Barnes pointed out in his comments on invention. Some people long for the nonexistent day when "the thing speaks for itself," but generally, technical people are realists. They know that as research organizations grow and research expenditures increase, critical sponsors must be informed of the work of the laboratory and of its contributions to the company if support is to continue. And hence the demand that the boss be a facile communicator, for they know that his performance as an "agent" will strongly influence the manner in which their work is regarded by the organization. If the boss doesn't understand the work, or is unable to "sell" its significance, their future is beclouded. If he *is* effective here, they know this will affect the extent to which their accomplishments are "picked up" by the organization and developed into useful contributions.

Technical people also expect their boss to be a provider of resources — of money, space, equipment, and manpower — to enable the group to prosper and grow. They are well aware that the resources of the laboratory are limited, and that in some fashion those resources must be allocated among the competing groups within the laboratory. The details of the manager's activities in this area are not highly visible and consequently are not well understood, but the *results* of effective performance are very apparent. Again, the fortunes of the group rest largely in the manager's hands, as his people see it. Interestingly enough, success in this area is not necessarily regarded as being related to technical competence. Possibly from ignorance, people attribute success in acquiring resources to their boss's skills in political infighting and to his aggressiveness. However he does it, the boss must "perform" in this area if he is to receive high marks. Furthermore, success in providing resources can compensate for other inadequacies, at least in the eyes of the men who work for him. Some of those who expressed greatest satisfaction with their boss said, in effect, "He has a knack for picking good people, providing them with resources to do their work, and leaving them alone." Others said, "You've got to give him credit, we've never lacked money to buy equipment, and we've got good facilities. He doesn't understand my work, but he doesn't interfere." So long as this happy situation exists, people apparently are tolerant of deficiencies that otherwise could become very disconcerting.

People recognize that a multitude of minutiae arise that impinge on the group and that could be extremely disruptive if not filtered out and disposed of. Letters must be answered, phone calls accepted, visitors have to be taken care of, ruffled feathers have to be smoothed, and misunderstandings have to be straightened out. The boss must perform all these tasks if the group is to continue to function, and frequently such performances require considerable finesse. Technical people assign this role to their bosses, and scientists and engineers appreciate the boss's sacrifice in their behalf and express amazement that he is able to tolerate such an existence.

The highly personal viewpoints presented above do not represent a comprehensive picture of the work of a manager of research or development, but they do emphasize certain aspects of his work that loom large in the eyes of technical people. The most striking theme, it seems to me, is that the manager's role is perceived as a *facilitative* one. He helps provide the resources and establishes the environment in which the really important work — the research — can proceed expeditiously, in accordance with the creativeness and effort that the technical people can bring to bear on it.

In this role, the manager is not represented as the man at the center of the action. The technical people are the featured performers. The manager is in the supporting cast. One could argue that this representation may be no more than a rationalization, to cut a powerful figure down to size, so that the individual can feel a greater sense of adequacy in dealing with his boss. (The terms manager and "paper shuffler" are sometimes used almost synonymously, usually with regret for the loss of a good scientist.) Nevertheless, recognition of the existence of this representation may help provide insight into the attitudes of technical people.

The sensitive nature of the relationship between a technical man and his boss is perhaps best exemplified by the problem of formulating the program. As noted earlier, this subject was strangely absent from many discussions, yet many managers would undoubtedly regard it as their key responsibility. This subject is both complex and delicate, because it lies directly athwart the individual's perception of his own ego. Competent managers know intuitively that this is an area in which they must tread softly.

The scientist's strong drive for autonomy has been noted many times. Inherent in this drive is a self-image of a person who has mastered the tools of his profession — the various bodies of principles and practices that are involved in performing R&D. Furthermore, the mature scientist prides himself on being in touch with his field, in knowing the areas that are regarded as fruitful for investigation, the knotty problems that are resisting attack, the current interests of the key figures in the field, etc. From this context and in the light of his own ideas and interests, he believes fundamentally that he is in the best position to determine what course will be the most productive for him to pursue in his own research.

Growth to this state of autonomy seems to be an important part of the maturation of a scientist or engineer. A number of people remarked that when they first joined the organization as young men they had looked on the manager as Herr Professor. They had expected him to have a powerful voice in the determination of the program and to provide guidance and stimulation to the members of the group. But as they acquired experience and competence, their growing self-confidence led them to modify their views. They came to believe that the initial role they had pictured for the manager was both undesirable *and* impossible. It was undesirable, from their point of view, because it required them to accept a passive, juvenile role for themselves as technical people, a role that was incompatible with their newly acquired self-image, as men able to chart their own course. And it was impossible, they felt, because the diversity of work being done under the manager's responsibility and the pressure of competing tasks made it impossible for any manager to provide technical guidance to all. Thus, it is not surprising that a mature scientist or engineer with a strong sense of his own adequacy does not consciously ascribe to his boss a powerful voice in determining what he will work on.

This problem is further compounded by the state of development of industrial R&D as a social institution. Though some companies, including my own, have been engaged for decades in R&D, the appearance of large organizations with elaborate structures, complex facilities, and significant resources is a recent phenomenon. Social institutions evolve their roles and relationships both internally and externally quite slowly. Attitudes with respect to the legitimate exercise of authority, to the acceptable use of power, and to the subtle relationships between people in organizationally structured roles are deeply rooted in our cultural heritage and are not easily made explicit. These attitudes are acquired imperceptibly as we grow up, get educated, and go to work. They evolve and become differentiated for different institutions and different roles over many years. The roles and relationships of people engaged in industrial R&D are relatively new and, consequently, are still being evolved.

To the man who is oriented toward industrial mores and

practices, it seems natural to apply to R&D the same roles and relationships that apply elsewhere in industry. The industry-oriented man can be puzzled and resentful that technical people resist when these roles and relationships are applied to R&D. Conversely, the man with a doctorate and with intense exposure to a strong academic culture tends to believe that the roles and relationships of *that* culture are most suitable for R&D. It is unlikely that either is correct.

This conflict will be resolved as we gradually evolve new roles and relationships that are specialized to the industrial R&D scene. In time, these will become legitimate through custom and usage. These new identities must evolve slowly, too slowly for the comfort of many people, through the efforts and experiences of many people who are struggling with the problem. The subject is, and should remain, a controversial one as this process of evolution proceeds.

Another theme that should be rather sobering to the thoughtful manager is the extent to which he is perceived as being in a position to help or harm his people, depending on his competence, good judgment, and managerial skill. Within the laboratory, his people are not party to many of the discussions and decisions that vitally affect them. The manager's role of agent for his people is one of the ethical imperatives of his job. Yet it is a role he can underestimate, or even ignore, in the face of unrelenting organizational pressures to which he is exposed — pressures for results. Furthermore, his group members are not the only people who regard him as their agent: Higher echelons in the organization undoubtedly regard him as *their* agent.

Another theme, not so apparent in the comments thus far, involves the general climate for interaction between managers and their people. I mentioned earlier that technical people have potent tools for influencing the way in which they will be managed. These tools can be used directly and indirectly. It appears that, to a great extent, their action is covert. In general, any open discussion of the managerial role, of relationships between managers and their people, of "how I should be managed," is considered taboo. Indeed, these subjects are not easy to discuss. In view of the evolutionary state of the R&D manager's role, this barrier to communication is indeed unfortunate. Both managers and research people have an interest in the nature of the roles and relationships that eventually emerge as stable configurations in industrial research. Furthermore, both have important contributions to make to the evolutionary process.

It is interesting that each frequently uses similar terminology to describe problems of the other. Technical people often comment that their boss is so buried in details that he has lost perspective. Conversely, managers sometimes complain that technical people are so narrowly specialized that they "can't see the larger picture in which the organization must operate." The existence of this barrier to communication, and of these feelings regarding the shortcomings of the other, indicates the strong need to initiate a dialogue.

Managers are in the best position to take the initiative in starting this dialogue. One man stated the problem quite clearly when he indicated reluctance to express himself on the subjects presented. He said, "I'm very loyal to this organization and I really should discuss my feelings on these subjects with my own manager first. But nobody has ever asked me these questions!" Managers *can* "ask the questions." And all concerned stand to benefit from the results.

Lowell W. Steele

Supervision

12

In both industry and government, a costly and dangerous trend is emerging, a trend toward a kind of management system whose pattern of operation is incompatible with that required for highly creative performance by scientific and engineering people. It is a trend toward tighter budgetary controls and tighter organization of the work. It is brought about by the impact of greater competition and, even more, by faulty information concerning the relative effectiveness of various approaches to cost reduction and improved performance.

I believe this is a dangerous trend because tighter controls, while appearing to provide desired results over a short-run period, may actually damage an organization over a longer span of time. I think we see evidence to support this in some of the earlier observations of Donald Schon and Carl Barnes. As a countermeasure, I am proposing a management system that enables scientists and engineers to function at their most creative level. This is a system that has evolved from an extensive program of research by the Institute for Social Research of the University of Michigan. You have already read of some of the results of this program in the two essays by Donald Pelz and Frank Andrews. From our studies at Michigan we have made a number of observations concerning the performance of technical people, some of which will sound familiar to you after the comments of my colleagues:

When communication within the organization is frequent, scientists and engineers are likely to perform better. You will remember, for example, that patent applications by technical personnel are higher in such organizations, and employee evaluations of one another are also higher.

Frequent communication with colleagues who think differently from one another also improves an organization's performance.

Scientists who do communicate with one another, but who are not conscious of a need for competent colleagues — who maintain an independent frame of mind — tend to be better performers than those who are more dependent.

Scientists and engineers who see their administrative chief rather often perform better than those who do not. And their performance is better still when they can also set their own technical goals, or at least have some influence on their chief in setting those goals. Best performance is observed when the scientist has high self-determination combined with free access to someone in authority.

The potentialities of younger subordinates are best developed by the supervisor who can maintain the difficult synthesis of close interest in the young man's work without domination of it. (Funds alone cannot guarantee achievement. The large, well-financed research organization that curtails the feeling of freedom — even unknowingly — risks the danger of inhibiting creativity.)

If the technical man's personal motivation is low — if he is not deeply involved in his work — it is not advisable to allow more than moderate self-determination in his work. If his motivation is high, then full self-determination leads to best results.

These observations point to an important general conclusion: Scientists and engineers are likely to be most creative when their supervision is such that they feel substantial freedom in their work — in selecting their problems and goals, in deciding on the approach to achievement, and in interpreting their data — and when they have frequent interaction with their superior. These findings are valid for the administration of basic research, developmental research, and engineering.

I realize that skepticism greets such a conclusion. To some, it will seem that the conclusion is obvious, needing no so-called research to prove its validity. Expressive of this skepticism is the research director who says: "It is unnecessary for anyone to prove to me that my people should feel free in their work. I already know that. And because I know it, my people do feel substantial freedom." To this person, I will repeat the warning of David Allison, at the beginning of this section on the technical environment: Beware of that undue sense of self-aggrandizement! I will suggest, further, that you note the inconsistent opinions expressed in our chart, which suggest that the true feelings of a manager's subordinates are probably quite different from what the manager believes those opinions to be:

INCONSISTENT OPINIONS:	SUPERVISORS . . . Asked of supervisors: "How do you give recognition for good work done by employees in your work group?" Frequency with which supervisors say "very often":	EMPLOYEES Asked of employees: "How does your superior give recognition for good work done by employees in your work group?" Frequency with which employees say "very often":
"Gives privileges"	52%	14%
"Gives more responsibility"	48	10
"Gives a pat on the back"	82	13
"Gives sincere and thorough praise"	80	14
"Trains for better jobs"	64	9
"Gives more interesting work"	51	5

Many supervisory people are not aware of such discrepancies, feeling that their subordinates see things as they do. And yet such substantial discrepancies often do exist, as we have learned in the course of our studies at the Institute for Social Research. The information shown in the chart comes from a study of a large utility by Floyd C. Mann, one of our research people. Measurements of employee attitudes, as illustrated here and as embodied in Morris Tanenbaum's questions shown earlier, are essential in the development of the kind of management system I shall go on to describe in this essay.

Another kind of skepticism also comes up quite often when we recite the conclusions we derive from our research. This involves a more traditional bias — and I know you have heard it many times: If you allow too much freedom, people will take advantage and the organization will suffer. Our studies have proved quite the reverse of this, but they have also shown (and this may account for some of the persistence of the traditional view) that it takes time to institute such a program. Thus, during a short period — say, one year — the kind of program proposed here is likely to prove less productive than one that is oriented around tighter controls. We have a near-perfect example of this in a recent study. Because of changes in the company's overall operation, it was necessary to cut off the test program after one year. Let us see what happened within the organization during that period.

The study covered 500 clerical employees in four parallel divisions. (The fact that these are not technical people does not affect the end result.) Each division was organized in the same way, used the same equipment, did exactly the same kind of work, and had employees of comparable aptitudes.

The four parallel divisions were assigned at random to two experimental programs. To each program was assigned one division that had been high in productivity and another that had been low in productivity. Then, during the year-long period of the experiment, productivity was measured week by week. Employee and supervisory attitudes, perceptions, motivations, and related variables were measured at the beginning of the programs, then again at the year's end.

In one of the programs, costs were reduced and productivity increased in the traditional way: Decisions were made at a high level, jobs were timed, standards were introduced, and workers were put under pressure to produce the work called for by the standards. The other program was managed in accordance with methods we advocate: Decision levels were pushed down the line, and employees were permitted to participate in decisions affecting their work; at each level of management, supervisors had greater freedom of action within stated policy; managers and supervisors endeavored to achieve a relatively high level of participation in all activities and decisions, except those involving compensation.

What was the result? The divisions managed in a traditional way achieved a 25 percent increase in productivity, a result of orders from the general manager to reduce staff by that amount and of management pressure to produce a substantial increase in production. The two other divisions, on the other hand, in which subordinates were involved in decisions, showed only a 20 percent increase in productivity. Here too there had been a reduction in the size of the work group — but the workers themselves took part in the decision to reduce staff, being aware that management sought to increase productivity.

Given these results, any management might reasonably conclude that the traditional system is superior, since it produced about one-fourth more improvement. But would this be a prudent conclusion?

The only data normally available to management is the type we have just examined: data on productivity, earnings, costs. But the results of the one-year experiment look quite different when the above data are supplemented with a whole new class of measurements, dealing with the degree to which each division is functioning as an effective human organization. These measurements reflect the quality of the internal state of the division: the extent to which its members accept its goals and seek wholeheartedly to implement them, the confidence and trust subordinates have in one another and in their superiors, the adequacy and accuracy of communication, and the motivations and capacities for effective interaction and sound decision-making.

Looking at the two programs in the one-year study, and measuring each in terms of these qualities, we see significantly different results. To be sure, the traditional program scored higher on productivity, but this gain was made at the expense of the organization's human quality, for there were adverse shifts in such factors as desire to do one's job well, the feeling of responsibility for the work and production goals, the attitudes toward supervision, and confidence and trust in management. Employees felt that their superiors were relying more on rank and authority to get the work done at the end of the year than was the case at the beginning of the year. And the deterioration in the organization's morale showed up in still other ways: Turnover increased; employees began to quit, saying they felt excessive pressure for production. (Toward the end of the year, when evidence of this began to appear in exit interviews, the company lessened the pressure somewhat.)

Quite the opposite was true in the other program. When increased participation was provided, there was an increase in the employees' feeling of responsibility to see that the work got done; more favorable attitudes were developed toward efforts to increase production. When the supervisor was away, employees kept on working. Employee attitudes toward the manager, assistant manager, and supervisor became more favorable.

The point of this example is simply that changes in costs and productivity show up rapidly when direct hierarchical pressure is applied, as, for example, when jobs are placed under standards. On the other hand, when participative management is introduced, improvement appears first in the attitudes, communication, and motivation, and later in productivity and costs.

What if these programs had run longer? When would the new system of management have begun to show significant advantages in terms of higher productivity and lower costs? Probably not for another six months. In another experiment recently completed in a service industry, six months after the introduction of participative management evidence was available indicating a favorable change in attitudinal and motivational dimensions. Significant improvement in productivity and costs, however, did not occur until a year and a half after the change in management practices.

There can be no doubt that, in the one-year experiment referred to earlier, the traditional system, with its inherent hierarchical pressure for higher production, was creating forces within itself that would soon have adversely affected its higher level of productivity.

I know this to be the case because of a substantial and growing body of research. Studies done at Michigan and elsewhere over the past two decades show clearly that, in terms of probability, the managers who achieve the highest productivity and lowest cost operations are those who, more than the average, involve their people in decisions related to their work and to their roles in the organization. This research involves hundreds of studies in virtually every kind of industry and occupation, and in many parts of the world: Western Europe, Japan, India, as well as the United States.

The findings from these studies indicate that the general pattern of operation of the highest producing managers differs from that of the managers of mediocre and low-producing units in the following ways:

First, we find in the highest producing managers a preponderance of favorable attitudes on the part of each member of the organization toward all other members, as well as toward superiors, the organization, and all aspects of the job. These favorable attitudes toward others reflect a mutual confidence and trust. To be sure, these are not attitudes of easy complacency; rather, they are attitudes of identification with the organization and its objectives, as well as a high sense of involvement in achieving them. Second, this highly motivated, cooperative orientation toward the organization's objectives is achieved by harnessing all of the many major motivational forces that can exercise significant influence within an organizational setting, and that can be accompanied by cooperative and favorable attitudes. Reliance is not placed solely — or even fundamentally — on the economic motive of buying a man's time; nor are control and authority the organizing and coordinating principles of the organization. On the contrary, all the following motives are tapped so as to yield cumulative and reinforcing influences and favorable attitudes:

The ego motives — including the desire for growth and achievement in terms of one's own values and goals, as well as the desire for status, recognition, approval, acceptance, and so on.

The security motive.

Curiosity, creativity, and the desire for new experiences.

Economic motives. These are used fully and effectively, but not to achieve control and authority.

By tapping all the motives that yield favorable and cooperative attitudes, we achieve maximum motivation oriented toward the goals of the organization. In part, this is accomplished because the needs of all members of the organization are fulfilled. And at the same time, we thwart the substantial decrements in motivational forces that occur when powerful motives are pulling in opposite directions — forces that exist when hostile and resentful attitudes are present.

A third characteristic we find in the most highly productive managers is that their organizations consist of tightly knit, effectively functioning social systems. This system is made up of interlocking work groups that have high group loyalty among the members and favorable attitudes and trust between superiors and subordinates. Also present are sensitivity to others and relatively high levels of skill in interpersonal interaction and group functioning. These skills are what permit the effective participation in decisions on common problems of which I spoke earlier. Participation is used, for example, to establish organizational objectives that are a satisfactory integration of the needs and desires of all members of the organization and of persons functionally related to it. Communication is efficient and effective. There is a flow from one part of the organization to another of all the relevant information important for each decision and action that is taken.

Finally, we find that measurements of organizational performance are used in high-producing groups primarily for self-guidance, both by the superior and his staff, rather than for superimposed control. To tap the motives that yield cooperative and favorable attitudes, rather than hostile ones, participation and involvement in decisions is a habitual part of the leadership process. This kind of decision-making, of course, calls for the full sharing of the measurements and information that are available. Moreover, as it becomes evident in the decision-making process that additional measurements or information are needed, steps are taken to obtain them.

In achieving operations that are more often characterized by the above pattern of highly cooperative, well-coordinated activity, the highest producing managers do not fail to use the technical resources of the classical theories of management — resources such as budgeting and financial controls, even time-and-motion studies. But the high-producing managers use these resources in a quite different way, and with greater effectiveness. This difference arises from differences in the kinds of motives that the high- and low-producing managers feel are important in influencing human behavior.

The low-producing managers feel — as do many of those using the traditional theories of management — that the way to motivate and direct behavior is to buy a man's time and then exercise control through authority. In keeping with this concept, the various technical resources are used by these managers to perform functions such as organizing the job, setting standards, and establishing performance goals and budgets. The decisions arising from the performance of these and similar functions are then superimposed by authority on the organization. Compliance is sought through the use of hierarchical pressure.

The highest producing managers feel, generally, that this manner of using technical resources does not produce the best results, that the resentments created by this direct exercise of authority tend to limit its effectiveness. These managers have learned that better results can be achieved when a different motivational process is employed. They seek to use all those major motives that have the potentiality — when properly used — of yielding favorable and cooperative attitudes. They try to use these motives in such a way that favorable attitudes are elicited and motivational forces are thereby established — all of which are mutually reinforcing. For example, their leadership is such that the motivational forces that stem from the economic motive are not blunted by employee collusion created by hostility that restricts the quantity or quality of output, because the full strength of all motives — economic, ego, and other — is used in a cumulative manner.

This description of the pattern of management that is more often characteristic of the high-producing rather than the low-producing managers points to what appears to be a key or critical difference. The high-producing managers have developed their organizations into highly coordinated, highly motivated, cooperative social systems. Under their leadership, the different motivational forces in each member of the organization have coalesced into a strong force aimed at accomplishing the mutually established objectives of the organization. This general pattern of highly motivated, cooperative members of the organization seems to be a central characteristic of the management system being used by the highest producing managers.

It is possible to derive and state a general theory of organization based on the integration of the management principles used by the highest producing managers. Obviously, the new system is a more complex form of human organization and thus requires greater skill and understanding in interacting with other persons.

Particularly important to people who manage the work of technical professionals is the evidence that this same management system is much more in harmony than are the classical systems with those principles of research administration that are associated with high scientific and engineering creativity.

For example, if we compare the newer system of management with the more traditional ones, we find that the newer system:

Provides more accurate and effective communication in all directions — upward, downward, and laterally.

Enables subordinates to exercise more influence on decisions affecting them and their work, while at the same time enabling superiors to achieve greater coordination of effort.

Achieves higher levels of confidence and trust among all members of the organization.

Enables members of the organization to work with a greater

feeling of freedom and with less sense of unreasonable pressure from superiors.

Establishes an organization in which there is less tension, less anxiety.

Yields greater productivity, lower costs, and less waste.

To shift an organization from a traditional system of management to one that involves a high level of employee participation is neither easy to do nor can it be done rapidly: Human patterns must be changed and these changes must be reasonably well synchronized between a superior and his subordinates; otherwise, misunderstandings and conflicts in roles will likely become excessive.

One early step that can be used to facilitate a gradual shift to a participative system involves the introduction of periodic measurement of such things as employee attitudes, motivations, and the adequacy and accuracy of internal communication. As superiors and subordinates gain experience in interpreting and applying the results of such measurements — and in observing the nature of the improvements achieved — they acquire progressively greater skill in participative management.

The leaders of industry and government are now free to choose between these two systems of management. Their choices will have a profound effect upon the creativity and motivation within their organizations — and, indeed, upon the destinies of their organizations.

It seems unfortunate to me to reflect that the choice in many such instances will be to stand pat, to preserve the traditional system. Indeed, traditional management may be likely to argue: "Why should the research department be managed any differently from accounting or production? This company cannot have one set of rules for research and another for everybody else." The point is, of course, that *all* elements of an organization benefit from the kind of participative management advocated here. If a degree of participative management exists already in the research department, this is to the good; other departments of the organization would benefit from emulation, rather than compelling research to conform with tighter budgetary controls and greater hierarchical pressure for increased productivity. Such conformity makes the role of those who direct engineering and scientific staffs increasingly difficult.

Of all United States corporations, those which are making the most extensive use of R&D and engineering are, on the average, keeping under accurate surveillance, by their present account procedures, the smallest proportion of the total value of their corporation. If total value is computed as total number of shares times the current price per share, the balance sheets of these firms typically keep under careful accounting scrutiny from about one-sixth to one-third of their current total value. Under such accounting controls, managers can easily report as earnings cash increases that are not earnings but merely liquidation of human resources caused by the kinds of high-pressure management discussed previously.

To provide a corporation with reasonably accurate information on the state and trends in the value of its human organization, customer loyalty, and similar assets, a program of research to develop human resource accounting is now under way. When developed, these new accounting procedures will enable a firm to make reasonably accurate estimates of the current value of these human assets. A corporation will be able to tell when an increase in cash represents a true increase in earnings and when it does not. Moreover, when managers build a more productive human organization, this will be recognized, even though it may be a few years before its actual value becomes manifest through the kinds of current earnings that are so carefully accounted at present.

When the leaders of our great corporations begin to insist on obtaining really adequate and complete measurements of the performance of their organizations, including human resource accounting, I believe there will be a dramatic shift away from the traditional systems of management. Periodic measurements of psychological and organizational variables—such as attitudes, motivations, communication—in addition to measurements of productivity and costs, should be used until human resource accounting is available. More accurate measurements of organizational performance will one day lead us toward the newer management system, and toward an administration of research that enables the scientist and engineer to achieve significantly greater creativity than is the case today.

Rensis Likert

The Organization Part III

Science and the Organization 13

One of the paradoxes of the technical world is the innocence with which its own mythology is accepted. For all the searching for new knowledge, or truth, which is regarded as the mission of the scientist, and for all the efforts at putting that knowledge to useful application — the function of the technologist — there is scarcely any thought that perhaps these same skills and methods might be used to appraise the myths of science and technology. If you have followed these essays in sequence, you have already encountered several bits of the mythology: Carl Barnes' citation of the better mousetrap myth, for example, and Donald Schon's account of the so-called rational view of innovation. On these next pages, you will encounter still others, myths that are clung to with the same devotion, despite the perilous foundations on which they must rely: the myth that holds that the success of Bell Telephone Laboratories is due to its autonomy; and the portrait of the omniscient technical manager, by John Hoskins, who is said to run his organization by uncanny rules of thumb; and perhaps the most deceptive of them all, the myth that describes the ritual by which the U.S. government evaluates technical proposals, then — fair and square — bestows largess on those that have won support in open competition.

It is perhaps understandable that such myths should be, for it is easier to believe that rules of thumb really work, or that Isaac Newton's great insight was due to a falling apple, than to have to contend with more complicated information, even though it is closer to the truth. But it is harder to know why we operate from the mythology, why we pretend it to be truth when in fact we know it is not, or suspect it is not. Perhaps we do this because we want to believe the myth, or, having nothing to take its place, because we feel we must believe, lest we find we have traded the myth for a sack of uncertainty that leaves us powerless to act.

The essays that follow are dedicated to the proposition that complicated methodology is more valuable than oversimplified mythology, and to the attainment of a better accommodation than now exists between the technical world and the industrial organization. Why should one wish for a "better accommodation" between the technical world and the industrial organization? So that private enterprise can "tap the reservoir of science" and thereby control its markets and maximize its profits? Some may see these as sufficient reasons, but I doubt that such goals are particularly attractive to scientists, including those who would work in the industrial environment. I believe there is a much more important reason for striving for a better relationship between these two: The industrial organization that does not establish a good relationship with its technical people cannot succeed at innovation. Such an organization is forced to be imitative, fearful of risk, skeptical about new ideas. Indeed, this is probably a fair description of many industrial organizations and may account for the fact that so many important technical ideas of this century did not occur within organizations, but rather were the ideas of individuals who had no large organizations to support their work. The jet engine is one such example, and others include the helicopter, xerography, rocketry, frequency modulation, Edwin Land's "instant" camera, and many more.

Undoubtedly, these examples are sources of embarrassment to those industrial enterprises wherein such ideas might have occurred — and most certainly to those who were offered such ideas by their inventors, but who turned them down as "impractical" or of limited potential. These missed opportunities on the part of large organizations are sometimes cited as indictments against the kind of system within which large organizations can flourish: If the giant corporations are unwilling or unable to foster the generation of new ideas, so the argument goes, why should such giants exist at all? Following such reasoning, it is suggested either that the giant corporations be broken into smaller pieces or that they be absorbed into a "seminationalized" structure under federal control. But neither course would promote innovation. In fact, either alternative would quite likely result in a reduction of innovation. What is ignored by those who advocate such alternatives is the essential role that can *only* be played by the large organization: the role that combines technical knowledge, capital, and capable management, and that culminates (when the process is working) in the attainment of something new and socially valuable, which otherwise would not exist. This is the *total* process of innovation of which the idea is only the catalyst. To point out that many large organizations are not very good at making this process work, however true that observation may be, does not prove that one or another of the alternative organizations would be any better. All it really suggests, it seems to me, is that most large industrial organizations lack a working knowledge of the innovation process — which is to say, a lack of understanding of the complex relationships along the chain that extends from research to the marketplace. This should come as no surprise, for the innovation process is one of those few activities of the technical world that has suffered in recent years from lack of support for serious study. Edward B. Roberts, whose essay on government support of research appears later on, has commented on this irony: "The government annually spends billions on research and development activities yet probably devotes less than one million dollars to research on R&D. Major corporations devote millions and even hundreds of millions to R&D on their projects and processes, yet most spend nothing on 'research on research.' Such backward attitudes result in the present situation; only a few organized groups in the United States are conducting broadly based studies relevant to the management of R&D."

What Roberts is hoping for in his call for more "research

on research" is an improvement in the innovation process. And certainly that is a worthwhile objective. But another benefit would come from such studies that could be just as valuable: Were we to know more about the process of innovation, we would appreciate why large industrial organizations are so essential to that process. For example, the lore of nylon: You will recall the story of Wallace Carouthers and the research he performed, while still at Harvard, on condensation polymers, and how the Du Pont company invested less than a million dollars in Carouthers' work when it was ready to build its first pilot plant for nylon. Quite true. But what must also be said of this achievement is that this was not the story's conclusion, for it would be more than a decade, and only after the investment of some $27 million, before the first dollar would be earned from a bolt of nylon fabric. The transistor and the 360 computer series are more recent examples, and just as striking in terms of the importance of the organizational role: The transistor could conceivably have been developed earlier; certain of the fundamental properties of semiconductors were known as early as 1885; it was known in 1904 that semiconductor rectifiers were well suited to detect radio waves. Why, then, did it take so long for this development to occur? Because there was a lack of knowledge of semiconductor science and because no organization appeared to develop that knowledge; only after the Bell Telephone Laboratories took up the development, which involved a large research gamble despite that earlier work, and only after an investment exceeding that for nylon, did the transistor become a reality. The transistor effect itself was a new physical reality in 1947, but it took until the early sixties before the promise of the field of solid-state electronics caused Bell Laboratories to invest more than $200 million in continuing exploration. A gamble of like proportions was made by IBM when it went forward with the development of its 360 computer. Would any of these developments have occurred without the support and capabilities of such large integrated industrial enterprises? The assumption is hazardous. Indeed, as time passes and the development stakes grow larger, the words of David Lilienthal become all the more discerning; nearly a generation ago, Lilienthal wrote: "Only large enterprises are able to sink the formidable sums of money required to develop basic new departures; a small corporation is rarely able to risk those large sums, perhaps enough to wreck the company if the gamble fails, on the success or failure of a major new project in such areas as electronics or chemicals, for example." Indeed, money alone cannot account for it; there must also be specialized people, joined together in a common purpose. And there must also be *time,* long periods of time, for that purpose to be achieved.

But if it is to be the large organization that will perform the task, what sort of organization will that be? For example, will it be the kind of organization we have constructed to put a man on the moon — publicly funded and federally controlled? If so, then we will have abandoned or demoted the kind of organization that developed modern electronics in favor of the kind that made the space program possible and the kind that has since been commissioned to create a supersonic transport. This will confirm the prediction of John Kenneth Galbraith, who says that "men will look back in amusement at the pretense that once caused people to refer to General Dynamics and North American Aviation and AT&T as *private* business." Indeed, the *private* portion of the first two of these giant corporations is now so small as to constitute only minor parts of their total activity; for the most part, both General Dynamics and North American Aviation serve but a single market — the federal government — and in this sense they are already the kind of organization Galbraith seems to favor: "If the mature corporation is recognized to be part of the penumbra of the state, it will be more strongly in the service of social goals." One must wonder at this. Can it be that Boeing, Republic Aviation, and General Dynamics — each of whom depends upon the federal government for most of its income and hence is already "part of the penumbra of the state" — are truly more socially conscious than other enterprises, e.g., IBM or AT&T, that are less dependent upon government support? One must ask: What is their record of contribution to public interest? But there is another reason for questioning whether the kind of "mature corporation" that Galbraith foresees would indeed be a stronger social servant. For example, he suggests it would be healthy to recognize that the telephone company is already a part of the nationalized complex that includes North American Aviation, General Dynamics, and the other government contractors; I suggest it would be healthier to recognize that this is *not* the case, that AT&T is still, in fact, *private* business and that, for precisely this reason, it has been able to be an innovative organization.

Yet the Galbraith prediction may hold and men may look back, not in amusement but in chagrin, at this "pretense." If the prediction holds for AT&T, and for other huge enterprises that he does not specify but that might include IBM, Du Pont, and others, then an important quality might well have passed out of existence — the innovative quality that those giant corporations provide. Bigness alone is no guarantee that this quality will exist, but bigness is one of the prerequisites, as you will see in Jack Goldman's later account of the basic research program within the Ford Motor Company. And when this is coupled to an understanding of the innovation process — when technical talent, capital, and management capability are brought together as the components of what Morton calls a "people process" — then a powerful force for change is set in motion. As you will see in Morton's second essay, on the current change taking place in the electronics industry, this force has shortened the time span for innovation in electronics from 50 years (for vacuum-tube technology to develop) to 20 years (for transistors) and now, perhaps, to 10 years for the development of microcircuits.

This story of change in electronics is really two stories in one: On the surface, it is the story of the revolution that goes on today in a single industry. This is a story of rapid and disruptive technical change — change that comes about as engineers and scientists begin to learn to make a great step forward, from making one device at a time (a transistor) to making whole circuits at a time. The second story is a kind of microcosmic movie of our times, a movie that rushes along at great speed, like the Keystone Cops. Morton never really tells this second story in his essay; rather, he is counting on our experience and imagination to reveal it to us. He is saying: "Here is a picture of the electronics industry as the pace of technological change increases. This industry has just recently emerged from one revolution, brought on by the transistor, and it is now experiencing a second revolution — the microelectronics dilemma. These will be painful years for many people in electronics, fateful years for many companies, but those who learn to survive and grow in this new era of microelectronics will enjoy the rewards of this new technology."

But the kind of change that Morton describes is happening elsewhere — in industries outside electronics and, indeed, in our society as a whole. Perhaps the only real difference between the electronics revolutions and others is the *rate* at which each change occurs. For example, we think of our urban revolution as one of rapid change — and the same for the current world dilemma over population — but actually the rate of change within electronics has occurred much faster. The electronics industry was completely transformed during the 20 years of the transistor revolution, and the second revolution, now occurring with the advent of integrated circuits, may run its course even faster and be even more disruptive than the first. Hence, within electronics' shrunken time span we see the *kind* of change already occurring elsewhere. For example, the people of the aircraft industry will see the relevance of the microelectronics dilemma. Morton talks of the systems companies in the electronics industry — such as the builders of giant computers — and how these companies feel compelled to extend their technical province to include new capabilities in the science of the solid state. The aircraft industry is witnessing a similar development among the systems companies in aircraft, with the manufacturers of airframes trying to extend their capabilities to include the components that make up the aircraft.

If the story of microelectronics is akin to the story of this age in which we live, Morton's account of the two electronics revolutions is a useful model for study. Let us think of his account as a speeded-up movie of our times and ask: What does the movie tell us? I believe it tells us two things. First, we see how an industry has been transformed twice by new technology based on science. On the first round, with the exploitation of the transistor, it became possible to produce smaller, more reliable solid-state devices at about the same cost as vacuum tubes. On the second round, in this current era of microcircuits, it appears it will be possible to produce whole electronic circuits at about the cost of producing a single transistor, vintage late 1950's.

Second, this fantastic change in electronics suggests what might happen in other industries, and in society generally, if the electronics model has relevance elsewhere. For example, suppose that something like the electronics revolutions had happened in the automobile industry. Perhaps the first development — the transistor analog — would have been a new power source, costing and weighing about one-tenth as much as the eight-cylinder engine of the early fifties. And perhaps the second development — the microelectronics analog — would be spurring the automobile industry today to work toward the production of an entirely new kind of vehicle, one that would be produced in a single step, an integrated auto, with unit costs dropping to a minute fraction of current costs.

As one examines the various patterns evoked by Morton's essay, the possibilities multiply. For instance, he speaks of the need in electronics for many kinds of technical specialists to deepen their knowledge in their various specialties, and he stresses the importance of coupling between specialist disciplines for a common purpose, so that a new level of overall systems capability can be achieved. Thinking now of the urban revolution, is it not also true that our urban centers have urgent need for this same kind of specialization and coupling? To raise the level of our urban society, we must use the talents of specialists — technical specialists, behavioral specialists, and many more — and these specialists must be moving toward a common purpose. It is significant and heartening that many urban professionals are looking upon their complicated problem in this "systems way." The city *is* a system, after all, a system of *people*. To be sure, its very existence is made possible by various technological developments: the steel skeleton, the elevator, vehicles of mass transport, and so on. But what has often been ignored as our cities have grown is the *reason* for their existence. Somewhere amid this confusion, the cities' *raison d'etre* was thought to be to serve the technologies that made cities possible, rather than the people for whom those technologies were meant. This is an error that systems thinking can correct. Indeed, this may be one of the most important contributions that modern science and technology can provide an urban society.

There is another kind of contribution that science and technology may provide. To the society at large, it might well be imperceptible, for this contribution occurs within the corporate organization — and there it is seldom regarded as a contribution at all, for it is too disruptive to merit corporate sanction. I am thinking of the power of science to dig into the *structure* of a corporate organization. This is different from the power of science to cause the organization to set off in a new direction, prompted by a new discovery and the potential profit envisioned with the discovery's exploitation. This is a newer and more subtle kind of power, not readily understood by all those who run corporations, for it seems to oppose the corporation's primary objectives of expanding its existing markets and improving its profit position. The scientist in the corporate organization may hope his ideas are marketable and profitable, but his personal objective goes beyond this: His professional goal is to make some contribution to science and to society, and he sees his corporation not as a mechanism for profit maximization, but as a means by which he can reach his own objective. This is the nature of the scientist's power — and the nature of the corporate dilemma. It is a bewildering and frustrating condition to many corporate managers, but to those who understand the problem, it is seen as a means by which the corporation can achieve a new level of accomplishment. To these corporate managers, the scientists are not demanding any denial of the profit system. What they are demanding is simply an answer to a question: "What is this company for?" The corporate manager who appreciates the connection between science and corporate growth knows that the answer to the question must be something more than "profit." He knows that the corporation's objective, like the scientist's, has to have more to it than this if the corporation is to continue to keep the best scientists on the premises.

But if science has this power to dig into the structure of an organization, the corporation has a countervailing power. It has the power simply to ignore the question or to declare anew that "the name of the game is profit." And, obligingly, the scientists will stop digging. But this does not stop science. It only stops the digging in that particular place, for the scientists move on. The exodus is comforting to some corporations. Life is peaceful now. But it can provide no comfort to the corporation whose future is tied to the exploitation of science, nor to the corporation that wishes to be a leader in tomorrow's world. To exploit science, to be a leader in the society of tomorrow, the industrial corporation must be able to attract the brightest young people from the colleges and universities. A generation ago, this was not so serious a challenge to the corporation, for the brightest students had fewer choices among possible careers. But such is not the case today, for the young scientist no longer must decide between a couple of attractive possibilities — perhaps one in an industrial laboratory and the other on a college faculty. Today he can decide among many possibilities, some of which may be in industry. What is his choice likely to be? If he is a typical member of this new generation, he will choose the job that offers high social value, the position that will enable him to work on technical problems whose solutions will be of value to society: helping to solve the world food problem, for example, for the environmental pollution problem, or one of dozens more.

Will he find this opportunity in an industrial environment? Not necessarily. He will not look for it in an industrial organization whose only objective is to produce new products at a profit. Rather, he will seek the organization that is more than a product producer and whose objective goes beyond profit maximization. This will be the organization with great strength in the sciences, but it will also be an organization that applies this excellence, in the products and services it provides, to the fulfillment of social needs. This is not a denial of the profit system, for profit is an essential component in the kind of organization to which I refer, just as return on investment is also essential. The point is that these should not be the *only* reasons for the organization's existence; rather, they are means by which the organization measures its effectiveness and creates new capital; that other reason — phrased in terms of "What is this organization for?" — completes the description of the kind of organization I believe will be attractive to young scientists and engineers.

One must concede that too few such organizations exist, certainly too few to deal with the kinds of contemporary problems that beset our society. As Harvey Brooks says: "One of the central issues of our time is how to deal with our pressing social problems, the problems brought about by the growth of population, urbanization, and the rapid application and diffusion of technology itself. These are public problems. They represent needs that cannot currently be expressed in terms of a market demand that can be satisfied for somebody's profit. There is no lack of ideas for dealing with many of these problems, but there is nothing analogous to the pull of the market to induce the development of the solutions, or to do the sorting out of alternative innovations that is achieved more or less automatically through the probing of the market in the private sector." Brooks believes that more of our thinking and "innovative energy" should be aimed at *inventing* a market for these public sector problems: a "social mechanism that has some built-in automatic selectivity that rejects unsuccessful ideas and rewards successful ones."

What form might such a social mechanism take? Jerome Wiesner has used the analogy of a giant learning machine, a machine that is constantly receiving feedback from its tentative innovations. To be sure, this is the kind of mechanism that exists now to provide our *private* goods and services. The new "learning machine" must be a similar mechanism, which is sensitive to the *public* market. This means it must be capable of hearing error signals from that market, in terms of profit fluctuations, and it must receive sufficient revenue from the market to enable the mechanism to sustain itself and grow. As Brooks says, until such a set of institutions is invented for our urgent public problems, it is unlikely that the "push of ideas" from research will be sufficient to sustain a continuing innovative process.

When one looks with optimism into the years ahead, I believe one can say with some certainty that science and technology will be major forces in shaping the world. If we succeed in solving such large contemporary problems as population, pollution, and the many others, the solutions will have to come via the efforts of technical people. But what is less clear, and hence must be cause for caution, is whether these technical people will belong to corporate organizations. The forecast is hazardous because we do not know whether the corporations will *want* to be involved with such problems. Nor do we know whether they will be able to organize themselves to deal with problems of such magnitude. And indeed, even though some may be eager and able, it is not clear that they will be permitted to do so.

The corporation whose sole objective is profit is not likely to venture far from its principal skill — the development and production of profitable items. Hence we cannot look here for much more than new hardware. The more venturesome corporation, perhaps wishing to create a new transportation system or an entire urban region, will find it quite difficult to identify in advance the eventual profit from such a venture. Hence this corporation, though wishing to extend itself, may not be able to take the risk. It may take on the task under government contract, employing scientists and engineers to carry it out, but now it is no longer a private enterprise. Rather, it is a government contractor and the technical people are government employees who happen to work under the contractor's roof. And finally, if the corporation does have some objective beyond profit, and if it has the resources within itself to take on a great task, plus the courage to risk failure, still it cannot be certain it will be allowed to carry out its purpose. A case in point is the Bell System's experience with communications satellites. Bell had invested some $75 million in its Telstar project, plus many times that amount in research and development that had general application to satellite communications. The objective was to meet an expanding market with a more effective mix of technologies. It is argued that it was in the public interest for the federal government to establish a wholly new organization outside the Bell System — the Comsat Corporation — to exploit this development, but at the same time one can appreciate the innovators' dismay when their research investment was canceled before it could provide new services and income. Bell does own 29 percent of the total shares of Comsat, but one must wonder still whether these innovators will be willing soon again to risk such a huge commitment to a technological advance, or whether they must now be more cautious.

If caution is to be the course that the large industrial organizations must follow, if the major technological risks can henceforth be taken only under government contract, as "part of the penumbra of the state," then it may be true that the finest hour of technological innovation will not come tomorrow. It will have come yesterday.

David Allison

14 Basic Research in Industry

A few years ago, on a visit to the Soviet Union, I had the opportunity to spend a most stimulating day with one of the most remarkable scientists of this generation, Professor Lev Landau. I was taken in to meet him, and before we so much as said hello, he said to me—in fluent American-accented English — "Tell me, what is Overhauser doing in a motor company?" He was asking about Albert Overhauser, a brilliant theoretical physicist on our staff at the Ford Scientific Laboratory. The "Overhauser Effect" was well known to him and Overhauser's affiliation with Ford was something he had learned from reading the *Physical Reivew*.

But what was Overhauser doing at Ford? Why is Ford doing basic research? Why is research an important ingredient in any modern industrial complex?

It is now more than ten years since I left the academic world to join the Ford Motor Company and help in the establishment of a basic research program in Ford's then-new laboratory. At the time, I had been teaching physics and doing some research at Carnegie Tech. When my friends in the scientific community heard of my impending move, they had a single reaction: "You are going to hit your head against a stone wall. You will never attract any first-rate scientists, either to the Midwest or to an automobile company."

Now, a decade later, we have a respectable and respected laboratory. We have about one hundred PhD's, at least half of whom are among the top men in their respective fields, and out of their endeavors have come not only good new science, but also significant impact on technology.

However, my purpose in writing this is not to extol my colleagues. Rather, one purpose is to answer a question: "Why should there be any basic research in the auto industry?"

At least five significant reasons can be given for carrying on research in industry. One is purely defensive: If you want to stay in business at a time when technology is marching rapidly, you cannot risk the chance that somebody else will beat you to the new technology. You have to be at the frontier yourself. If somebody, somewhere, is about to make a major contribution — a contribution that could obsolete your product — you must have people who will know about this, who will know what is happening. If your people are truly productive scientists, original scientists, creative scientists, then they will be sought out. They will be invited to give seminars. They will invite their friends to give seminars. They will attend meetings and conferences and be in touch with the scientific community. They will know what is going on. It is not very likely that an important development will take place somewhere in the world without their knowing about it.

If you are to tap the world's science and technology, you have to create some science. Your admission ticket to the club is to have something of your own to talk about.

The second reason for doing research comes right out of the first: If you have those good people — if you are at the frontier — you maximize the probability of doing some of the innovating yourself. This clearly is a competitive advantage.

Third, research is an increasingly important source of management personnel. Many of today's best college people are youngsters who want to go into research. This is true in science, in engineering, in economics, and so on — and this is an excellent way to get people into a company. Start them off on the research road. Top management today needs technically oriented people and one of the best means of finding them and keeping them is through the research lab.

The fourth reason is that research provides you with possibilities for diversification. A high-quality research program will invariably lead into new fields. It can even create those new fields. Moreover, the company that is amenable to diversification can the better justify intensive research areas and programs that may be too costly to justify on the basis of relevance to existing business alone.

There is yet a fifth reason: what I call, in poker parlance, the "keep them honest" reason. Many industrial enterprises maintain extensive in-house technical capability — embracing product, design, process, development, and manufacturing engineers. These technical people are largely confined in their efforts to proprietary company-oriented material, with little or no contact with the outside engineering community. How do you evaluate them? How do you measure them against the standards of society? By maintaining a basic research capability, a company is assured of at least a minimum complement of technical people who are constantly evaluated by their professional peers and, therefore, subject to the most critical and objective evaluation mechanism available. This group then provides a standard against which the balance of the technical capability can be measured.

Having offered several reasons for doing research, and having skipped others—for there are many—the following question comes to mind: How do you maximize the productivity of a research enterprise? How do you best utilize it to benefit the company?

People in research management live by a faith, a conviction. The faith I live by is this: "What is good for science must ultimately yield benefits to Ford." Not "good indirectly." I do not mean, "If we benefit science we benefit the company." I mean this: "Put together the best possible people. They will thrive and do good research. And, inevitably, as sure as night follows day, out of this will come something that could benefit the company."

Now this does not give you license to support *all* science. You support those fields that relate to your business. Ours is transportation, and this tells us to be interested in energy —

energy as a resource, the conversion of energy, the storage of energy, the utilization of energy. It also tells us to watch electrochemistry, photochemistry, semiconductors.

We are also in the materials business, and this takes us into metallurgy, solid-state physics, polymer chemistry. And we are in the devices business: Electromagnetic and optical devices are important now, and we can look ahead and see their importance increasing. This part is easy — just as it is easy to see that high-energy nuclear physics and steroid chemistry are *less* relatable to transportation than are these other fields.

The initial delimitation of fields — at least in the broader sense — helps the odds of success a little. At least the research effort has been prevented from dissipating in all directions. What can be done beyond that? How can the probability of research "payoff" be enhanced without demeaning the concept of "basicity"?

The analogy I like to use is poker. After all, every world — including the world of physics — is statistical. You play the laws of probability. To maximize your winnings, you must recognize that there are three important ingredients. One is plain, simple luck. If you are not going to get any cards, you are not going to win.

Coupled with luck is the skill of understanding probablity: how to use probability to your advantage; how to take advantage of the odds. And finally, there is the intangible ingredient of psychology: assessing what the other person is thinking . . . what he is doing. The good poker player can influence what happens out of all proportion to the ordinary laws of statistics.

Management science — research management — is really a gigantic poker game. You employ these three ingredients to try to maximize your gain. To begin with, there is luck. The legendary apple hit Sir Isaac on the head. Had he not been under that tree, had he not had that inspiration at that moment, etc. But many people have been hit by apples. So luck alone will not do the trick. In research, at any rate, luck or serendipity may be less important than understanding the odds — the odds of people and the odds of ideas. For example, what is the probability that a given area of research will be successful — measuring success either in terms of scientific or economic yield?

Suppose that the problem is the development of an economic fuel cell, a fuel cell that will be useful in the transportation field. One must ask: What is the likelihood of this problem's being solved? We can deploy this many forces and that many facilities — but what are the odds? Now, many times the problem will not be undertaken — research will not be done — because you come to the conclusion that no matter how much money you spend to achieve the payoff, you will not be able to earn enough, even if your results are successful. So you hold back. But let us say that the problem we are considering looks solvable, looks promising. It is still a gamble — still a poker game — but how do we take advantage of the odds?

Let's look again at the fuel cell, and specifically at this question: What sort of effort should there be in fuel cells? Should we mount a big general program? When you study this question, you find that, in the present perspective of things, fuel cells are not yet in the ball park — whether in terms of economics, energy density, or power density — for personal transportation. For spacecraft, the economic base is different, but for personal transportation you come to the conclusion that a major effort of an applied nature is not yet called for.

But then you ask yourself a second question: If somebody, someplace, is going to make an important discovery in fuel cells, where and how is this likely to happen? You are forced to recognize that this will probably come out of the catalyst, because the development that will spell the difference in fuel cells — both in economics and practicability — will probably happen in the catalyst: Find a cheap, simple catalyst and fuel cells may come into the picture.

So you decide on the following course: You mount a first-class program in understanding fundamentals of catalysis. This is meaningful basic work for an industrial laboratory. It may involve fundamental research in electrochemistry, in solid-state physics, in radiation effects or diffusion. If you start out by getting top-notch people, you have bought two things: knowledge of what's going on in the world, and a weighting of the odds in your favor that innovation will take place within your own organization. From your efforts, you certainly will have learned much about the problem — a greater corporate asset than the ill-advised publicity you get by jumping prematurely into applications. In our laboratory, this approach has indeed paid off. We staffed just this sort of program and out came not a fuel cell at all, but a completely new and exciting concept in batteries, the Ford sulfur battery.

When a new laboratory is established, recruitment is tough. In our early days, people would lift their eyebrows: "Research in the automobile industry?" They thought of this industry as reactionary, unreceptive to new ideas. There was only one way to overcome this. We had to get people who were recognized in science. Competition was terribly keen then, when we were trying to get the ball rolling, but the next step was easy.

Once you get the first few, and others see that your professions of scientific grandeur are not mere idle protestations, others are eager to come. Two of the key men in the lab came here simply because another man, whom they respected highly, was already on our staff. This has always been a major consideration among scientific people in choosing a habitat, whether on campus or in industry. It was important to establish such an image quickly by finding a few such men, and this we set out to do.

There is a synergistic effect in research. The whole is certainly greater than the sum of its parts. A laboratory made up of interacting compatible greats is more likely to yield impressive advances, both scientific and practical. And one must not minimize the role played in this process by the unsung stimulus whose name may never be attached to the published paper or to the invention. Without him, the critical step may never have been achieved. But in any event, you must hold out for top-notch people; if they are not obtainable, for one reason or another, it is better to leave slots unfilled and budgeted money unspent than to delude yourself into thinking that you are covering a significant field adequately.

This is, in fact, the approach we took in such fields as lasers and cryogenics, which we considered vital in terms of long-range potential to science and to our company. The results have been rewarding to both.

But no industrial research organization can do this unless it has the interest and understanding of top management. If you look at organizations that have tried research and failed, you find — almost invariably — that top management had no clear idea of what research really was, what it was for. Conversely, those industrial organizations that are eminently successful in creating and utilizing good research invariably have top managements whose enlightened dedication to intellectual progressivism is well recognized. At Ford, for example, it was top management that initiated the now-celebrated five-year plan for funding its Scientific Laboratory.

Having allowed that research is a necessary ingredient for industrial security and growth, what are the prevailing considerations that enable the optimization of research productivity and maximize the impact of research on industrial goals? I choose to classify the requirements into three major categories, which I label, respectively, atmosphere, leadership, and coupling.

Atmosphere: Since research is the company's liaison with the future, it must be freed of all the traditional forces that inhibit the free thinking of competent minds. Atmosphere is intended to include the many freedoms that competent research personnel require in order to render their minds fertile and their actions conducive to creativity and innovation. It means the freedom to allow one's mind to wander across uncharted fields, freedom to go off on tangents, freedom for people to interact with each other and with their co-professionals in other institutions and other lands. And it must provide a system of rewards and motivations so that the scientist or engineer can receive just compensation even though his accomplishments may not be so tangible or comprehensible as those of the salesman or manager. And it must make adequate provision for facilities to support the ambitions and requirements of a highly skilled technical staff. It is as imprudent to be overextended in personnel and delimited in facilities support as it is to be overfacilitated at the expense of capable support personnel.

Leadership: The successful research enterprises have consistently attracted and built their organizations around recognized intellectual leaders. A roster of the known and successful industrial research enterprises invariably recalls to mind the greats of science and engineering who not only provided the magnetism to attract high-grade personnel but constituted the essential stimuli and criticality needed to make a loosely knit body of scientists and engineers function as an interdisciplinary unit. It is not sufficient to legislate into existence a research enterprise and set it up as an institution to gather knowledge at the frontiers or apply such knowledge to new materials, methods, or devices, staff it with recruits, and hope that it will go its merry way. Any institution, whatever its goals or charter, thrives only when there is vigorous leadership that can chart its goals and give substance and appreciation to the aspirations of its personnel. Moreover, in today's milieu it is the only proven mechanism of attracting competent younger men.

Coupling: Finally, there is the key (but often neglected) aspect of modern industrial research: coupling science to technology, coupling the goals of the individual to the goals of the corporation, the problem Lowell Steele described when he talked about "the man in the middle" — the technical manager who must try to resolve the problem. This is a very subtle concept. It recognizes that the goals of the individual scientist or engineer may not necessarily be those of the corporation; that the individual scientist (and this is particularly true of basic research) worships his intellectual peers wherever they may be. His audience is his fellow scientists. It is their approval and approbation that he seeks.

How can these ambitions be reconciled with those of the profit-seeking corporation? This is the challenge of research management. There are many ways by which this can be accomplished; which is to be chosen depends on many specifics that depend on the people and the corporate goals. Sometimes the motivation to recognize corporate problems comes from within. Sometimes it must be catalyzed from without. Sometimes it is provided by a team of liaison men or research planners; sometimes it is provided by the research managers themselves. But it must be provided. For research to have economic impact, this coupling must be provided. Many a research institution has been an unrewarding asset to an industry or a nation for want of a strong coupling mechanism to weld the scientific life of the nation to the technological needs. This in no way implies or suggests that the purity of science will be diminished or its goals altered; it merely means that science cannot just stand by itself if its fallout is to be properly exploited.

Jack Morton talks about this coupling function and how it works at Bell Labs; his account follows this one. Morton talks in terms of the flow of ideas, and of *people,* from basic research to engineering, and so on. Our approach at Ford is similar to Bell's, although the organizational patterns are different, reflecting the different types of business and market orientations of the two enterprises. Whatever the nature of the enterprise, different components of the research endeavor will be responsive to different pressures and appeal to varying types of audience. The responsibility of each such group is clearly marked: A product research group "faces" the product, and its audience is the marketplace and the marketer; a manufacturing research group "faces" process technology. To a basic research laboratory, such as our Scientific Laboratory, the world of science is its audience — and its principal product is publication.

This is the way it must be, for otherwise the man in the Scientific Laboratory will not be a good scientist, and this is what you want him to be, first and foremost. This goes back to the faith recited at the outset. If you have developed a high-level community of scientists, not only will the fallout from their pure science impact on technology, but they provide a cadre of professional backing for the diverse requirements of a complex corporate organization. A top-notch theoretical physicist in the employ of an automotive company may owe his primary allegiance to his colleagues and peers in physics, but he is readily excited if his colleagues across the street seek his counsel on a hydrodynamic problem in combustion. Or take the case of a man in our Theoretical and Mathematical Sciences Department who is working in the field of general relativity. He is making significant contributions to that field. Long after he had been working almost exclusively in relativity, he found an important topological relationship between some of the space concepts he is evolving in his relativity work and some unusual kinds of configurations that came up in metal physics and in a particularly unusual mechanical linkage. It is a strange chain, but it is there — and the presence of this man on our staff, thinking about these topological relationships, is extremely useful in guiding some of our other people. He provides a point of contact with a field of science with which there might otherwise never be contact.

Now, this does not say that the scientist is encouraged to wander over into technology — or rewarded for doing so, or forced to do so. Good research has to be cushioned. Perhaps I phrase it best if I say that good research must be insulated, but not isolated. It has to be insulated, or cushioned, because once people learn that they can utilize this talent to put out fires, to help solve immediate problems, then the research is crippled.

An appropriate question to ask is: How do basic and applied research groups fit into a company's long-range planning efforts? It is precisely in this arena that the subtleties of motivating scientists toward corporate goals (when their instinctive loyalties and orientation are professional) receive their acid test. There are several organizational mechanisms for accomplishing this. One that we have found successful is to involve the scientific people in that phase of corporate committee structure with long-range implications. We have had a particularly exciting experience in the field of long-range transportation planning. Corporate management was concerned about future trends in transportation and what roles the company may play in the new transportation. How will it affect our product? What should the company attitude be toward the automation of highways? How will it affect the way we engineer our products? What about the future of energy sources? Or the availability of materials?

A corporate committee was created to conduct a long-range study. Scientists participated, side-by-side with economists, market researchers, engineers, product planners, and systems analysts. The resulting studies were given visibility at the highest level of corporate management. Some of the results: the emergence of a need for a full-time, high-level technical group to do transportation and planning, unrestricted and unconstricted by present product plans or market realities. (By locating this group organizationally in the Scientific Laboratory, it was possible to attract some of the best people in the field and provide a healthy and creative environment.) The Ford electric car program owes its origin, as well, to the interaction of the scientists, engineers, and economists within the framework of this corporate effort.

What I have outlined in the last few remarks has been one approach to this critical problem of coupling. Are we optimized? Only time can tell. There is no doubt in my own mind that in providing the ingredients of leadership and atmosphere we are on the right track toward optimization of research productivity. The "coupling" problem is, however, far more subtle. We think that, before an accurate assessment of our success is possible, there must be greater opportunity for organizational experiments that have been successfully tried and tested at other labs. Two important experiments that Morton will talk of, relative to Bell Labs, are the flow of people, and the existence of a systems engineering group. One cannot overemphasize the importance of having a cadre of scientific alumni who have moved from the basic research laboratory and taken key posts in those organizational groups that are more preoccupied with applications.

Many corporations shy away from basic research altogether

— on the assumption that this is better left to the universities. There is no question but that the universities play a dominant role in this national resource that we call basic research. But I do believe that a large, diversified organization, in the billion-dollar class — if it has some dependence on technical advance — can ill afford the impedance mismatch and long relaxation times necessarily entailed in a dependence on extra-house research. Its technical people need the close interactions with men who are active on the frontiers, which is only possible through geographic and organizational proximity.

Funding of research is another problem that must be carefully considered. There has been a growing reliance of industry on the government picking up the research tab. This is unfortunate; we have been lucky enough to avoid this pitfall. If somebody outside the corporation is funding the research, there are psychological and economic pressures to do the research you can sell, not necessarily the research you think is important. If basic research in industry is to survive and be useful, it must have some semblance of longevity and should not be affected by the cyclical ups and downs of corporate sales. In our laboratory, for instance, we operate on a five-year budget.

In a technologically oriented industry, it seems to me, this is the only way you can function. You must operate with long-range planning. You must plan the financing of the company, and its technological developments as well, so that technology keeps pace with the ever-expanding interests and involvments of the marketplace. When long-range planning is done, technology becomes interwound in the corporate structure — research becomes an accepted part of the operation of the company, and the company literally sits in the grandstand of science.

This cannot happen when the research activity is tied to the company's profit-and-loss position. You cannot put research onto a corporate balance sheet and satisfy the financial wizards of its worth. Of course, we could put together such a balance sheet if we were to recite all the accomplishments of science over the years, all the products of today's marketplace that did not exist three decades ago.

A few documented case studies may serve to illustrate more cogently this amplification factor in maximizing technological output from scientific input. The case of the transistor, amply demonstrated in Morton's upcoming account, is one celebrated example. In our own laboratory, some very basic experiments on surface wetting and interfacial energies between liquid metals and inorganic salts led directly to new cutting materials that are likely to revolutionize the tool-bit industry. And more recently we had an example where the diverse talents and capabilities of several people — one in magnetic resonance, one conversant with thin film techniques, one with wide experience in low-temperature phenomena — were fused into one in order to try out a single critical experiment — macroscopic quantization in superconductors. As a result, we think we may have opened up an entire new electronic technology based on the quantum mechanics of the one macroscopic system in which it can manifest itself in the laboratory. Or I might mention the work that some of my colleagues are doing in the field of nonlinear optics; like so much of the research in industrial laboratories that can boast of viable programs of basic research, this work doesn't appear on the surface to be related to automobiles. This is as it should be. For so long as industry's laboratories are producing good science, respected science — we know such work is helping to make our communication system better, and our transportation system better, and so on. But perhaps more importantly, these research efforts assure us that the telephone company will not go the way of the telegraph companies, and that the lot that befell the locomotive companies will not befall our company.

There are broader implications of research excellence on the national and world scene — implications that have been a source of personal stimulation and keen interest to me. To the microcosm of industry, the pursuit of excellence in research has been represented as being vital to competitive survival and economic growth in a technologically oriented milieu. The logic is readily extrapolatable to the national and international scene. This suggests that the in-house laboratory in the government must pursue, establish, and maintain the same level of breadth, interdisciplinariness, and excellence that is a *sine qua non* for the successful industrial laboratory. It is a matter of record that, where this has been carried out in deed, the consequences have been rewarding and, not indirectly, the civilian economy has benefited as well — because once again the adage is proven that good research will pay off — although you don't know at the start where it will pay off.

In a like vein, developing countries on the international scene seek a type of economic growth similar to that of the competitive industrial organization. It might be well to contemplate, therefore, how a typical country, given certain resources, markets, population capability, can best deploy its technical talents in a basic and applied research enterprise directed toward a set of unique goals established for itself. This certainly offers a challenge not unlike the one offered me when I approached the task of developing research programs within the automobile industry.

Jack E. Goldman

15 From Research to Technology

I want to talk about the process whereby we convert relevant basic research into new technology. And I stress the word *relevant,* because the range of science is so broad that some areas are more easily converted than others, on a given time scale. As managers of research organizations, we are interested both in understanding this process and in improving its efficiency. Actually, we are in the same position as the engineer or scientist who wants to improve the efficiency of a machine, or who wants to develop a better one: He must first understand how the machine works.

The method of understanding I shall talk about, relating it to research management, is precisely the same method the scientist uses to understand how a molecule works. The molecule is his machine. He knows it works. He can observe its properties. But he wants to understand why it behaves as it does, for this understanding will enable him to do things with it, perhaps even to design new molecules that are better. The scientist calls his process the scientific method. He uses it because it disciplines him to ask the right questions at the right time, innovate at the right time, and tells him when to test those innovations.

The scientist follows certain steps in his process, and I shall take these up in turn before going into a discussion of the research management process. You will see later on that the management process is a direct steal from the scientific method. What are these steps? First is a clear statement of the problem the scientist is trying to solve. And, incidentally,

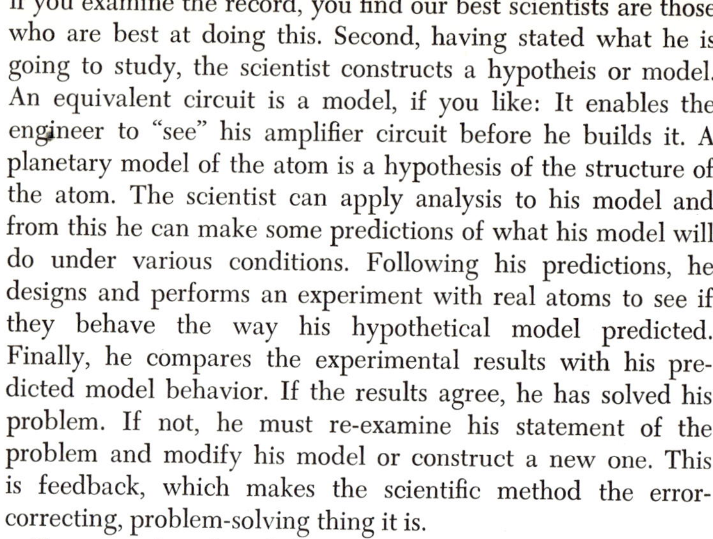

if you examine the record, you find our best scientists are those who are best at doing this. Second, having stated what he is going to study, the scientist constructs a hypotheis or model. An equivalent circuit is a model, if you like: It enables the engineer to "see" his amplifier circuit before he builds it. A planetary model of the atom is a hypothesis of the structure of the atom. The scientist can apply analysis to his model and from this he can make some predictions of what his model will do under various conditions. Following his predictions, he designs and performs an experiment with real atoms to see if they behave the way his hypothetical model predicted. Finally, he compares the experimental results with his predicted model behavior. If the results agree, he has solved his problem. If not, he must re-examine his statement of the problem and modify his model or construct a new one. This is feedback, which makes the scientific method the error-correcting, problem-solving thing it is.

For example, when Bohr imagined the planetary model of the atom, he followed these same steps and found that, indeed, many predictions agreed with real nature. What had to be added to his model were the exclusion principle and the quantum effects; these led to a new model, namely, the Schrödinger-Dirac quantum model, and to the quantum picture of the atom that exists today. The important thing about the scientific method is that the model need not be perfect on the first guess. Eventually, it will get better — the series will converge, as the mathematicians say.

Now to our problem: We want to understand the process of converting relevant research to new technology. This is the statement of our problem, our first step. Next we want to construct a model of this process. And as we do this we must recognize that this must be a model of a total system, all parts of which are related and important. In other words, our process is not just a flash of genius, not just the discovery of a new piece of understanding, not just the development of a new product or a new piece of manufacturing equipment. Rather, the process is all these things, acting together in an integrated way. And as we construct our model we see that it is really a model of how people work together, because our process is a people process: We are studying people and the structure in which they are related to one another. So our model, if it is to lead us anywhere, must help us understand what motivates these people — what makes Sammy run? And we must understand this in terms of their intellectual capabilities, their relationships with one another, their abilities to communicate. Not only must our model lead to an understanding of those things we know have worked in the past; it must also tell us of the failures. If it is a good model, it will enable us to build better systems of people, and to achieve higher and more challenging goals.

Another name for this method is systems engineering. This is the name we give it in Bell Laboratories, and it is nothing but the application of the scientific method to engineering systems: such as a communications system, a weapons system, an airplane, a ship. Or a system of people, or a society.

The first step in the systems engineering process is a clear statement of objectives, a definition of the goals of the system: What is the system supposed to do? In a satellite communications system, for example, we state that it shall carry 600 channels; it shall have such-and-such a bandwidth; it shall have a signal-to-noise ratio of better than so-and-so, distortion of less than so-and-so; and to be economic, must have a certain life.

In stating objectives, or goals, you must not be frivolous. If the goal does not bear a real relationship to what is feasible, then you are finished before you start. This is one of the measures of a whole man — a man who is at peace with himself — and of a successful manager: How well does he appraise both the needs and the limitations versus the potential? How good, how realistic, is his judgment?

The next step is to make as comprehensive a list as possible of alternative solutions. Even the simplest problem has more than one solution, whether it be a problem in physics or engineering. And the number increases when one moves into sociology and economics. Here it is important that one be aware of the maximum number of possible solutions.

But how is the best solution determined? The manager must develop "measures of effectiveness." And there are many such measures. For instance, suppose that I am trying to develop a system of transcontinental communications. One measure of this system is: How good is its signal-to-noise ratio? But another is: How much does it cost per channel? The clever solution is that which does not optimize one and badly degrade another. There is a balance, a trade-off between these that the manager tries to achieve. Sometimes timing is a very important measure of effectiveness. Sometimes you cannot take the time to get the best technical solution. Getting there first may be far more important than waiting for the "mostest."

Hence, many factors go into measuring the relative value of alternative solutions, and you must be certain that you know what those factors are, because this is the time you decide whether to commit long-term research and costly development, and whether it will be worthwhile if successful. And you had better measure from many different viewpoints.

Let me give you an example. From the viewpoint of economical, nationwide communications, all-digit dialing is a must. It is a necessary part of the economic solution to the growth problems for communications service. Further, it is something you must grow into, because it involves millions of telephones and billions' worth of plant. All-digit dialing was logical and economically necessary. As you know, however, some telephone users saw the problem from a different point of view.

It helps the physicist or engineer to think more about these things — these measures of effectiveness and different viewpoints, including the economic and sociological — when he is dealing with an engineering system. When the scientist is doing science, he practices the method intuitively. But when he is involved with a system involving people — a more complex system — he may overlook certain important factors if he does not consciously and diligently follow all parts of the method.

Thus far, we have restated the first step: We have defined

our goals, listed the alternative solutions, and developed measures of effectiveness to choose the most likely ones for further study. Having made some selections, we construct models of the most promising solutions and try to predict how they will behave under various conditions. Finally, we come to the experiment. On the basis of our predictions and our measures, we are ready to put the best hypothesis to the test. Because we are managers, our experiment might be the formation of a new division or embarkation on a new enterprise. We are about to lay our money on the line.

This is where we come to synthesis, where we form something we did not have before. And when we have done this, our experiment has begun — our organization is running. At this point, we must do precisely what the scientist does: We must compare what we have with what we were trying to achieve. We must monitor our signals and, as quickly as we can, feed them back into a restatement of our problem and a modification of our model if necessary. The fact that the scientist does this explains why science is in constant flux. The organization that fails to do it — that fails to change with its changing environment and changing problems — becomes a static organization. And, generally, a static organization is on the road to decay.

What this amounts to, then, is simply a restatement of the scientific method into the systems approach. If I am an engineer, I want to use it on engineering systems. If I am a manager, I want to use it to set goals, to design and to run my organization. In other words, the model for the management job is the role of the systems engineer. And what I shall do now, having set down this foundation, is try to apply the systems approach to the process of getting from basic research to industrial technology.

What has to happen before you can get from one to the other? Well, one thing must happen: Somebody has to do some research. We can agree that the function of our basic research is to synthesize an understanding of the physical world. But we know that mere understanding will not produce the new technology we seek. There are other steps we must take: We must demonstrate that this understanding can lead to a new device, a new circuit or system relevant to our business. This is the output of applied research. And we must show that this device can be manufactured at a price that people are willing to pay, and that it will satisfy our requirements for reliability. New measures of effectiveness come in at each of these stages. By putting these steps in sequence, we can begin to isolate specialized functions that are part of this total process. If one step is missing, all others can be good and still the whole process will fail. Hence it is essential that we see these things as being linked with one another. We must think of this process as a machine — both a forward-acting machine and a feedback machine — and the thing being processed through this machine is information, not hardware . . . information that goes from one person to another. When we see it in this way, we see that its building blocks are people and that there must be communication among them — both forward-acting and feedback — if this machine is to do its job.

We can list some factors that affect communication. There is language, for example: If these people do not understand one another, they are not going to communicate. There is space: They may talk the same language, but if they are too far apart, communications will be difficult. There is the structure of the organization itself — chains of command and all that. And finally, there are the things that motivate people: Are there enough common motivations among all these people so they will want to communicate with one another?

When we think of our machine as a "people system" for processing information, we see that we can do things that will inhibit the flow of information, and other things that will encourage it. Just as in an electronic circuit, you use insulators, conductors, semiconductors, to build barriers and bonds to the flow of electrical information.

And sometimes you want to build both. I don't apologize for using an analogy, because the construction of a model is a process of analogy. It enables me to get a bird's-eye view of the process and helps me to see the importance of the relationships among the pieces of my model. For instance, I can see that if I allow the feedback loop from design or manufacture to basic research to get very strong, the feedback will stop the basic research. And it won't be long before I've lost my research and perhaps my research people. So we purposely put a barrier between manufacture and basic research — either a space barrier or an organizational barrier, maybe both.

But what do we have that makes these people want to work together, to communicate with one another? What does the physicist have in common with the engineer in manufacturing? They have a commonness of goals that are challenging to all. This is what makes the man in basic research want to work in an area where his understanding is relevant to the organization. This is what makes the applied research man eager to pick up an idea from basic research and pass it along . . . and so on down the line. We have basic researchers who could be working on nuclear problems, relativity, vitamins, or what have you, but it is the common communications goal that motivates them to relevance.

For example, what made Bill Shockley and his cohorts pick the semiconductor area? Well, it went like this — and in describing this episode I think I am describing the kind of system the manager can have, the kind of courage he has to have, and also the interplay between scientist and manager:

Mervin Kelly was aware of the limitations of vacuum-tube technology as long ago as before the Second World War. Kelly had been a scientist in the electron-tube field himself, and as head of Bell Labs it was important for him to look 10 to 15 years ahead and realize that if we had to rely on electron tubes we would not be able to develop the more capable future communications systems at prices anybody could afford. There were simply too many built-in physical limitations to this technology. When he discussed the problem with people like Shockley, he was brought up to date on the current state of understanding in other areas: In our own laboratories and elsewhere researchers were trying to understand why a semiconductor is what it is . . . why does it behave as it does? Why is it different from a metal or an insulator? Wilson had studied this in England, Davidoff and Joffé in Russia, Shottky in Germany, our own people here at Bell Labs. So there was an existing body of theoretical work on hand.

Coming back to the electron tube, the problem was this: Relays were cheap and long-lasting, but too slow. This was their basic limitation. On the other hand, electron tubes were very fast. All they did was move electrons — a very low mass — but to move those electrons you had to pay through the nose in the hot cathode and the high vacuum. You used power in gobs and the cathode wore out. So we said, "We want it to be fast, like the electron tube. The machines of the future must handle lots of information and lots of bandwidth, so it must be fast. But we cannot pay the penalty the electron tube demands."

We wanted to control those electrons. In a conductor there are many electrons, but you cannot control them. In an insulator there are none, and nothing you can do about it. But in a semiconductor, aha — there can be lots of electrons or not, depending on what you do to the semiconductor.

People had already stumbled across the phenomenon of rectification. You put the voltage one way and caused electron flow; you put it the other way and caused none. Shockley had surveyed the various areas in which one might look for electronic phenomena. The field of semiconductors looked the most promising — if one could understand the basic physics. You see, the theory of that day would not explain a copper-oxide rectifier: It would not explain it quantitatively, nor would it predict the appropriate direction of rectification.

But we did know certain things. For instance, we knew we had to move electrons in such a way that the basic structure of matter would not be altered — otherwise, we would have the same kind of problem we had with the hot cathode. We knew there were lots of electrons in solids and that they would move as we wanted them to — without altering the basic structure. So the real question was: How can we control those electrons?

The decision to explore semiconductors was made by Kelly, but bolstered by Shockley, based on a phenomenon of nature: If you put the voltage on one way, you had lots of electrons; if you put it the other way, you didn't. Kelly and Shockley were making a synapse between a knowledge of need and a knowledge of possibility. They had not the slightest idea what form this electronic amplifier would take. But they believed that an understanding of the mechanism of electron conduction in semiconductors could lead to the synthesis of an electronic device.

Nobody could put a quantitative number on the probability — in the way gamblers can calculate the odds in poker. All one could say was this: We had an excellent chance of achieving understanding, because quantum physics had been exercised sufficiently; we ought, at least, to be able to understand elemental semiconductors, like germanium and silicon. And note that we started with the simplest semiconductors, despite the fact that more complex semiconductors already existed as useful empirical artifacts — copper-oxide, nickel-oxide, a range of them. We picked the simplest because we were seeking understanding.

Why don't we talk odds, probabilities? Put it this way: You are virtually certain you can solve the problem if you are willing to spend enough man-years on it. So it becomes not a question of odds, but rather a question of costs. It becomes a management decision: Are you willing to make this investment? You use the old biological computer to evaluate the odds, pay your dough, and take your choice. This is where you are tested for experience, good judgment, and that phenomenon called "guts." This is the way the transistor was kicked off. A major corporate goal came down to a narrow area of relevant science. I have told this story to show the importance of tight integration in the system: Everybody must know what the overall goal is, so that within each man's area he can look for those solutions that are most relevant to the goal.

Why do people want to do this? Well, most of us want many things from life. We want fame, fortune, fun, freedom, other things . . . you can name them. And we see, if we are mature people, that we must put something in. Anyone who isn't a bit odd picks the goal he can be proud of — one he can tell his associates, his wife, and kids about. If the goal is challenging, and is such that he can identify with it, then he has no trouble picking the relevant job, the relevant area of research. Indeed, he is eager to do so, proud to do so. The societal as well as the technical nature of the corporate goal is extremely important.

I maintain that when one establishes the purposes of a laboratory — provided they are broad and challenging enough — he does not inhibit the freedom of the people who work there. They still have choices — still make choices — but they do this on a conscious basis, knowing whether or not their choices are relevant. Let me give an example. Back in the early days of transistors, after we had learned how to grow single-crystal germanium and silicon, Bill Pfann had a sweet idea on zone refining. Bill is one of our most creative metallurgists and he had developed his idea to the point where he could grow a single crystal. One day, he was telling Shockley and me about this and we said, "Bill, we'd like you to do something else. We don't need your idea." We tried to persuade him that there was another development he could make that would be more relevant. But Pfann said, "To hell with you guys. This is important." And he went ahead with his idea.

Well, it turned out that Bill Pfann was right. The point is that we were really disagreeing on relevance, not whether it was good research. Shockley and I were measuring relevance against specific transistor objectives. Pfann, being a metallurgist, was thinking of its relevance to materials problems — not just transistor materials, but on a broader scale. In the area of materials research, his judgment of relevance was better than ours.

If a scientist in a communications laboratory comes along and says he wants to do work on drugs, you may have to suggest that there are other places where he would fit in better. You don't do basic industrial research because you think it is fashionable. You do it because it is an important ingredient for survival. Putting it another way, you can give it a biological analogy: You cannot afford cancer in any organism. Wild, disorganized growth will kill the organism unless the organism gets rid of the growth. How often have we seen large, successful corporations fail because of overdiversification? This is wild growth, going where random research takes you, a form of corporate cancer.

We will agree, I believe, that in basic research, people should not be confined to narrow branches of science. It is therefore important that the goals of the corporation be such that research can cover a broad range. But if research tries to cover the entire universe of science, if it tries to convert all of science to new technology, then one finds he is no longer in an integrated business. In fact, he may not be in any business.

Does this mean the industrial researcher is restricted in the work he can do? Let us compare his range of choices with those of the university researcher. We can see that there must be some mechanism whereby all areas of science are covered. For this reason, the universities are important contributors to new understanding: The universities must have complete freedom of choice with regard to research problems aimed at broad understanding. They must be free to tackle the most critical problems holding up complete understanding of our world.

On the other hand, universities should not neglect this basic responsibility in order to do industrial research. In this sense, both university and industrial researchers have different goal-responsibilities and associated comparable freedoms. Indeed, the industrial researcher may actually have greater capabilities for carrying out his basic research, because of the facilities and support industry should provide in its relevant areas of science. But in any industrial laboratory, this matter of relevance and goals must always enter in.

The knowledge of what business you are in is not necessarily stultifying. It can and should be challenging if industrial goals are properly selected. I believe the fact that we were in the communications business increased the probability that the transistor would be invented at Bell Labs.

Let us look now at the design and development stage in our information processing machine. In applied research, the job is to demonstrate technical feasibility of a new artifact from new understanding. In development and design, we must now carry the development much farther in technical detail but — most importantly — come up with a design that meets not only its performance requirements but its economic objectives of cost and reliability as well. The output of this stage is a manufacturable design that meets all the performance and economic factors required — at the right time. This widening of requirements and measures increases the range of solutions, calling for new judgments and providing new challenges to the design engineers.

What sorts of barriers and bonds do we need in our organization in order that information will flow as we want it to, from one stage to another? We do not want maximum flow in both directions. We want a balance.

At every interface — between applied research and design-development, for instance, or between basic research and applied — the motivation shifts, but the language is the same. Thus, the applied researcher must understand the arrangement of atoms in germanium and the role of impurity content — and hence must understand the basic researcher when he talks about it. These two people have similar backgrounds: You'll find many PhD's in applied research as well as in basic. The main difference is motivation — a difference in what they want to do with their education and understanding.

What this says, then, is that the old-fashioned idea, held by some, that you have all your bright people in research and all your drones in engineering, has got to go. In fact, the stronger the group in engineering, the stronger and more free the research group will be to explore new relevant areas.

Now we want some feedback, so let us see how we get it from, say, applied to basic: We get it, in one way, with a space bond — people in applied and basic live in the same building. And we get it through a common language. But at the same time, we see that if applied people or engineering people can dictate what the basic research people do, they will kill the long-range basic research. So we need an organizational barrier: One man — Bill Baker — is head of all basic research; other men head up applied research and engineering. Our people are free to sell, to stimulate and motivate all they like. But my engineers and researchers, for example, cannot tell the basic researchers what to do. And conversely, the basic researcher who believes he has made an important discovery cannot order the applied research or engineering people to pursue it. So this organizational barrier provides freedom for basic research and freedom regarding what shall be developed.

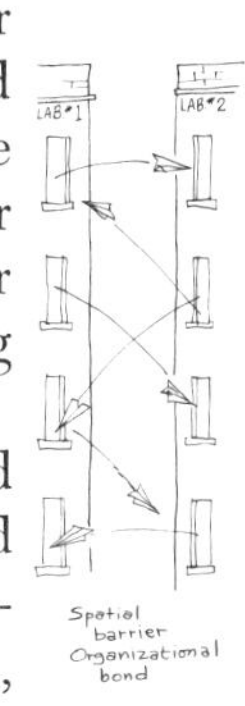

Spatial barrier
Organizational bond

What about the link between research development and manufacturing? Indeed, should there be a link, or should there be a barrier? In our organization, we say the laboratories are responsible for research, development, and design, and that manufacturing, distribution, and installation belong somewhere else — in our case, the Western Electric Company. Many years ago, a wise manager saw that a corporate barrier should exist between them. Otherwise, if both Bell Labs and Western reported to a common head too low in the system, a screaming emergency in manufacturing could cause all work in research to stop — the old story. Yet total separation is not the answer either, and I want to explain how we tackled this problem, because I believe it must trouble managers in many organizations.

Organizational barrier
Spatial bond

When we studied our machine as a system, one thing we began to understand was that it was bad to have both an organizational barrier and a space barrier between development and manufacturing. Together, these barriers make communication too difficult, and one has problems in transferring new scientific developments into new technology. So we built a space bond between the laboratory and manufacturing: We actually moved our development-and-design-for-manufacture group — a Bell Labs group — into a laboratory on the Western premises. Organizationally, they belong to the laboratory. But physically, they are linked to Western. And now we know we should never have a space barrier and an organizational barrier on top of one another. We use organizational and spatial links in complementary relations—wherever we have a space barrier we also have an organizational bond, and vice versa. This gives us an integrated design. And it works beautifully. Whether this is the kind of structure we will have two decades from now is doubtful. Technologies will change. And a different structuring may well evolve.

The prevailing myth tells us the secret of Bell Labs' success is due to its autonomy. The diagram above may account for the myth, for it is true that Bell Labs is separated organizationally from Western Electric, the recipient of the Labs' R&D. Bell and Western are linked via top management of AT&T, but each has its own president. In this sense, Bell Labs is autonomous.

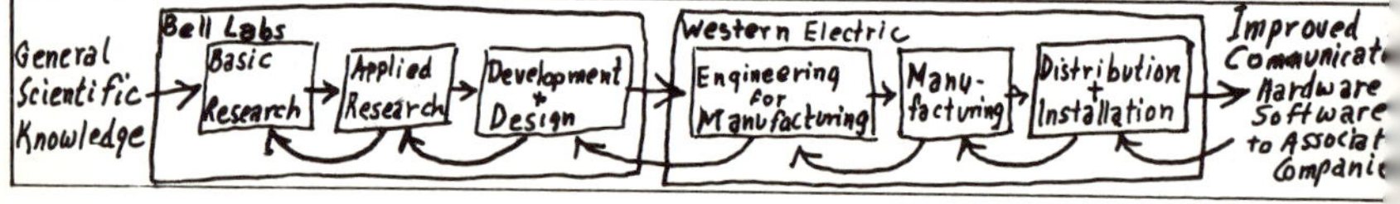

But Bell Labs is also part of an integrated system, as this flow chart shows. Author Morton describes the functions within the boxes as specialized parts of an information processing mechanism, with information flowing both forward and back. The specialized goals of each part—whether Basic Research or another—must be relevant to the overall goals of the integrated system. The main flow of information is forward, with enhancement and innovation added at each stage. The feedback flow represents appraisal and new needs. Feedback is the error-correcting evaluation of the main output of each stage.

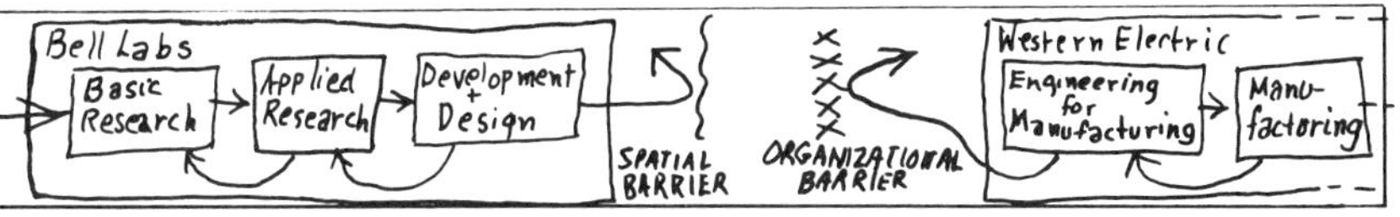

Until the end of the war, two barriers existed between Bell Labs and Western. One was organizational — each had its own president. This barrier still exists. The other was spatial — Bell Labs was physically separated from Western. But two barriers became a handicap to the forward flow of new designs and the feedback effect from manufacturing to design. Hence the spatial barrier was eliminated and replaced by a spatial bond, as is shown in the next diagram.

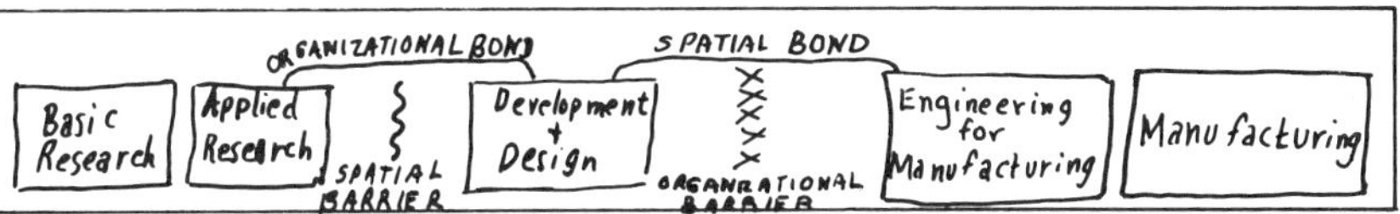

The spatial bond was created by moving the Labs' Development & Design groups onto Western premises. This bond enhances the day-to-day communications between Labs and Western engineers. Because a spatial barrier now exists between Applied Research and Development & Design, these two BTL functions are integrated organizationally within the Labs at the lowest possible level consistent with group size and their technologies. As this diagram shows — as well as the one below — a barrier of one kind is always accompanied by a bond of another. Further, two barriers never exist side by side, lest information flow be impeded.

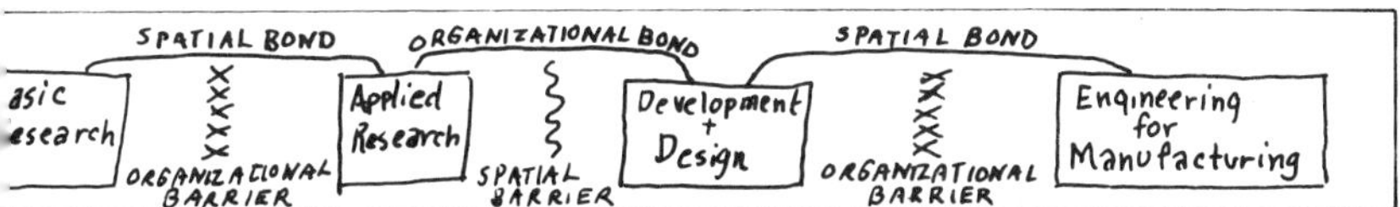

Spatially and organizationally, the specialized parts of the system are now linked together as seen here. An organizational barrier at the highest possible level (vice presidential) within the Labs protects the freedom and continuity of long-term basic research. Conversely, the fact that Basic Research and Applied Research are located on the same property provides a strong space bond between them. This bond aids flow of information and people, encourages dialogue on relevance of research areas.

If you will look at the chart at the right, you will see a separate laboratory function called Systems Engineering. Note that I have not put it in line: It is not a line function. About 500 people are involved here, under an executive vice president, Ken McKay.

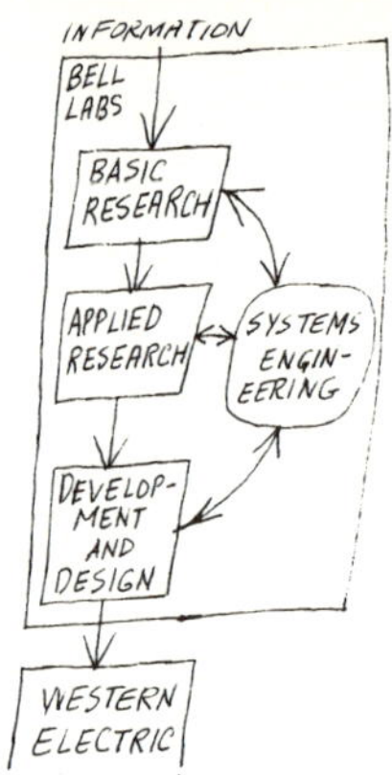

What does McKay do? What is the function of Systems Engineering? By training and practice, he is a physicist. By further experience, he is an excellent engineer as well. In Systems Engineering, he is concerned with economics and sociology, as well as with physics and engineering. One of his jobs is to build a bridge between the Laboratories and AT&T headquarters and the operating companies, and to be alert to needs and problems.

Because he is part of Bell Labs, McKay is also in touch with scientific-technical developments. From the synapse of these he can draw a plan for a proposed system. His study will say: If we succeed in doing the following, this new system will provide x millions of new revenue per year, or it will save x millions per year. His people will also take inventory of all the technology needed for success of the new system. Large-scale developments are not undertaken without knowledge of missing critical capabilities. Systems Engineering makes many such studies every year.

But McKay does not tell Baker he must do specific research, nor does he tell me to give him a specific gadget. He does tell us what is needed and what the priorities are, because the purpose of Systems Engineering is to provide overall guidance for research and development, in terms of corporate needs. Systems people are in constant touch with all groups and the realities of their needs and possibilities — operating company, long lines, research. This is the way we form our synapses. But the choice of exploratory projects in research is the responsibility of research people and their management, for they can make the best decision, given the fact that they know the needs. Each man, whatever his level, takes the overall goal, translates it into the goals for his range of science and technology, then makes his choice of relevance in those terms.

The systems method I have been talking about is applicable at any level in the organization. Let me show this by an example. Periodically, I sit in on management conferences at which we look at the needs of the Bell System and our various capabilities and programs. Baker is there, representing basic research. The other system vice presidents are also on hand. As I listen, one of the things that comes out clearly is that one of the big, important limitations on every system is that it is memory-limited. Whether we are trying to fix our Number Five Crossbar, our new electronic switching system, our data transmission, or whatever, most systems are limited in terms of the kinds of memory available.

Memory turns out to be a major technical and economic barrier to progress. Now I'm in the memory business: We are constantly searching areas of basic research for new concepts of how to store information, and how to get access to it.

So we conclude from these discussions, and from our own studies, that we must go for important memory advances — in a small way, just as Kelly, some 20 years ago, saw that amplification had to be done in another way than by vacuum-electron physics. I talk to my executive directors — the men who head semiconductors, ferromagnetics, ferroelectronics, ultrasonics, what have you. We agree, "This is a prime need." They go back to their lab directors, who go, in turn, to their department heads, their supervisors, their individual people. And they say, "We need better memories." Nobody says it's an order, or the boss wants it for his birthday. We give the reasons why and ask, "What are the present limitations?"

We hold symposia and in time it permeates the organization: "If anybody can think of any new phenomena, any new materials, any new technology that will make a major advance in memory . . . this will be an important contribution." It becomes a conscious, knowledgeable search. Each man down the line, within a more and more specific area, looks at his range of possibilities.

And it always happens: A number of ideas begin to come together. You always have more ideas than people to work on them. You have to make a choice: Here are the alternate solutions, the measures. Which ones do we push?

I believe this example is translatable to any business. But what is required? Take the rubber products engineer. What is the first thing he must know? He must know what the president of his company thinks is important: What are the goals? *And why?* And what does the vice president think is important, and so on down the line. Until he knows, he is in no position to choose which thing, from a range of things, he should do. This is the way to build a machine — an organization — that will allow information to flow and individual judgments to be made. The man must know, whatever his level, what the important goals are. And he must know not only *that* the boss said it was so, but *why*.

Conversely, it is that man's responsibility to transmit the things he is aware of, and what things are possible and relevant. Shockley supplied that function after he had heard Kelly sound off on the limitations of electron tubes. Shockley said: "I think semiconductor physics is the area to explore." This dialogue in goals must be carried on, up and down the line, deciding amongst alternate possibilities at each level: Which are relevant? Which might pay off? What will it cost?

The research part of any organization — big or small — has at least two functions. It generates relevant research. And it also recognizes relevant research, wherever it might have been done. For instance, the optical laser theory was originated by Townes, at Columbia; early experiments by Maiman at Hughes indicated one possibility. We got on board very fast because we were active in the field and we recognized its relevance to us.

Maybe the smaller research organization can generate only a little of its own basic research. Its research people must constantly fish the larger reservoir of science. But they must also contribute something to the general pool. You cannot have free access to the pool and understanding of its content unless you make a few contributions and are part of the research community.

If I have described a good, workable model in the forego-

The Evolution of a New Technology

BELL TELEPHONE LABORATORIES
INCORPORATED

CASE No. 38139
DATE AUTHORIZED
TO BE FILLED IN BY COMMERCIAL RELATIONS DEPT.
W.L. 7/16/45

AUTHORIZATION FOR WORK

SUBJECT Solid State Physics - The Fundamental Investigation of Conductors, Semiconductors, Dielectrics, Insulators, Piezoelectric and Magnetic Materials.

STATEMENT

Communication apparatus is dependent upon these materials for most of its functional properties. The research carried out under this case has as its purpose the obtaining of new knowledge that can be used in the development of completely new and improved components and apparatus elements of communication systems.

We have carried on research in all of these areas in the past. Large improvements in existing types of apparatus and completely new types have resulted. Thermistors, varistors and piezoelectric network elements are typical examples of new types. The quantum physics approach to structure of matter has brought about greatly increased understanding of solid state phenomena. The modern conception of the constitution of solids that has resulted indicates that there are great possibilities of producing new and useful properties by finding physical and chemical methods of controlling the arrangement and behavior of the atoms and electrons which compose solids.

APPROVAL
H. Fletcher J B Fisk 6/21/45

12.20-1	A. T. & T. Co.
CLASS OF WORK	Fundamental Studies Applying to Communications - General
COMMERCIAL RELATIONS MANAGER	

GENERAL DEPARTMENT HEAD
M. J. Kelly
Exec. Vice-PRESIDENT

E-770 (11-40)

"This is the document that initiated and authorized the basic research project which led directly to the invention of the transistor. The decision to explore semiconductors was made by Mervin Kelly, executive vice president of the Labs, bolstered by William Shockley, of basic research. This was 1945. Research and development had highlighted the needs and possibilities for this new approach to electronic components, and solid-state physics was selected as the area where basic understanding would be most relevant."

ing — if, indeed, this machine does work, then what it says to the man in a technology-dominated company is this:

First: know what business you are in, what its short- and long-term goals are.

Second: know your science and technology. Sharpen your relevance judgments. You cannot know the whole world of science.

Third: re-examine your goals continually, from the viewpoints of both needs and possibilities. An old idea may have been ahead of its time and is just the thing for today's need. Epitaxial semiconductor junctions are a case in point, as you will see in the example I call "The Evolution of a New Technology."

And fourth: know the meaning, the philosophy, and the methodology of systems engineering. This is not a process the Systems Engineering Department practices alone: Everybody's doing it. You do not have to be part of a formal systems engineering organization, but you should *practice* sys-

"By 1948 we had a device — the point-contact transistor. It was still a laboratory curiosity, though a curiosity of great promise. The physicists still had to understand and control the behavior of semiconductor materials, and this need led to development of a technique of growing single crystals of germanium. By 1951, with this knowledge in hand, we were able to produce a practical device — the grown junction transistor, shown here, which had been predicted by Shockley in 1949. Now we had both a new key to technological growth and a new basic tool for further scientific understanding."

tems engineering in running your own ring and relating it to the larger circus.

Let me conclude with a story to that point: In early 1948, the transistor had not yet been announced. The first "bailing-wire and sealing-wax" model had been shown to a few people. I had worked on semiconductors during the war, but by 1948 I was out of that area and was working as a supervisor on microwave tubes for the transcontinental radio relay system. One day Kelly called me to his office. He was

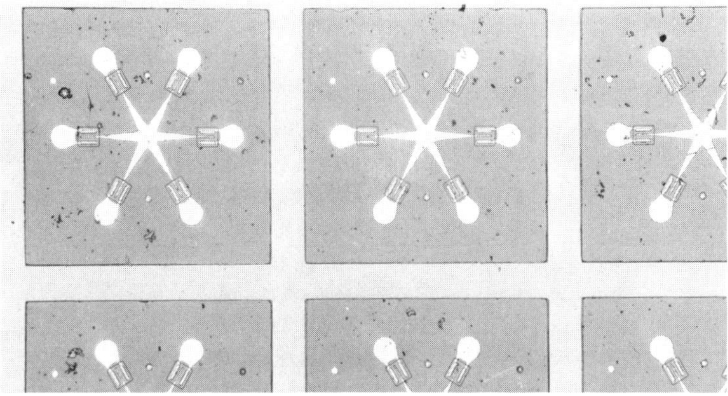

"By 1953, transistors could be produced in quantity and at a low price. This was the era of the alloy transistor, the most widely used device in the semiconductor family. The industry was now in a period of substantial growth. But the road never ends: Alloy technology also had limitations — it was a "one-at-a-time" process and, further, the frequency performance of alloy transistors was definitely limited. The search for solutions led us into the area of diffusion, which has given us multiple-unit production (see photo): We can produce 2000 of these six-transistor stars on a square inch of silicon. The search also led us back to an area we had explored before. This was epitaxy — a process of growing layers of material atop one another. Together, these developments allow us to build the elegant, balanced technology we need for integrated circuits and functional devices: This is the technology that will rid us of the tyranny of numbers, by allowing great reductions in the number of discrete devices we must design, fabricate, and interconnect one-at-a-time. Now we are talking of devices whose performance approaches intrinsic capabilities of semiconductor materials themselves."

executive vice president of Bell Labs then, and a formidable guy. He said, "Morton, I think you know something about transistors. . . . Don't you?" I said that I knew very little but that I realized they were pretty important. So he said, "We've got to follow through on applied research and development immediately. No time to waste. I'm going to Europe for the next month, Morton, and when I get back I'd like to see your recommendations as to how we should go about developing this thing. Good-bye."

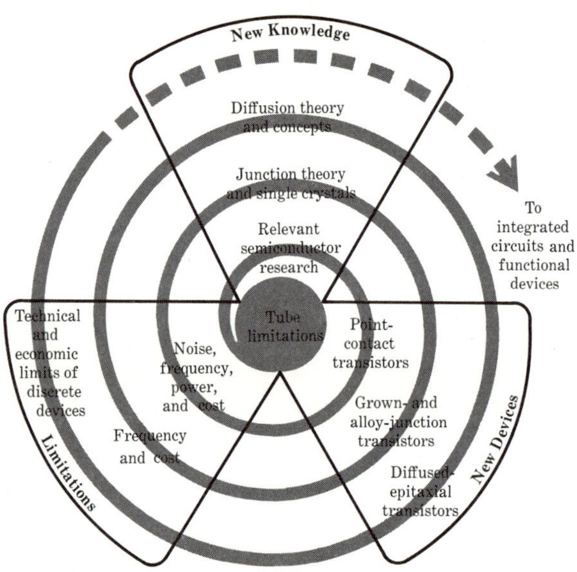

"Research and technology are not individual and autonomous, but rather are closely coupled parts of the total innovative process, which I have shown as a spiral. The first turn around the spiral begins with a clear recognition of the technical and economic needs of the present. Here one spotlights the relevant areas of basic research which provide the new concepts and materials for applied research which, in turn, provide the new device concepts and techniques for design, which, in turn, provide the input for the manufacturing phase. As these concepts and materials are developed to their intrinsic limits through this process, they also generate the new needs to be fed back into the machine and the second turn around the spiral begins."

For the next 29 days, I suffered torture. Instinctively, I thought I knew what to do, but I could not order my thoughts in such a way that I could lay out a plan, a system for what Kelly had asked for. It was not until the weekend before his return that I discovered how to get hold of it. I sat myself down and made my first conscious application of the systems approach to a program and an organization.

I do not say we developed the best possible program, but by using this best possible method — the systems method — we got a good solution: a program and an organization that could evolve, adapt, and grow in capability.

Jack A. Morton

16 From Materials to Systems

More and more industries are discovering that technological innovation is a major means to growth in economic capability. But many enterprises limit their innovation to new products, when it is the enterprise itself that may be most in need of change and renewal.

For technology to grow in capability, its systems must grow in complexity. To innovate more complex material systems, our "people systems" also must grow in complexity. Each must achieve increasing specialization and more intricate couplings of its differentiated parts.

The electronics industry is a good example of such growth in all its dimensions. Its relatively short history divides nicely into three eras of markedly different complexity, each characterized by the new electronic technology which gave it birth. These three steps are illustrated on the following page.

Each new technology was stimulated by the limitations of the old and the potential of new research. The large photo illustrates the one that started with the invention of the tube and reached its economic limitations in the form of discrete tubes and associated components, made one at a time, assembled and wired one at a time with hand tools.

The second wave of technology, represented (top right) by the logic circuit photo, reached its economic maturity using miniaturized transistors, diodes, and passive components — still assembled automatically one at a time but connected into a circuit by batch-solder techniques, on printed wiring boards.

Finally, the current microcircuit technology, still young though vigorous, is illustrated in one of its forms in the pair of photos at the bottom of the page. It is a synthesis of modern silicon-transistor and tantalum-film techniques in which complex circuits are made at one time from just two basic but compatible materials technologies. It has yet to reach its full technical-economic maturity but it is having major impacts upon the electronics industry.

The evolution of a new technology always provides opportunities for innovation — for renewal and growth in the capability of an industry. To profit from them, we must see and solve the problems that are unique to the new technology. These new problems always involve new science and technology and this was the case for tubes and transistors. As our technologies increase in complexity, the problems become increasingly people and management problems as well. Microcircuit technology is today's example. At the same time that it generates opportunities greater than before, it poses new people problems, which look tougher than the technical problems.

Technology of the solid state is derived from a broader, deeper field of science that was new and strange to most of us. New kinds of specialists, processes, and facilities in modern physics, chemistry, and metallurgy had to be developed and coupled together in new ways for the mastery of semiconductor technology. Some of today's electronics companies did not exist fifteen years ago, while others have completely changed in content and structure.

Microelectronics uses materials, structures, and techniques similar to those of transistor technology. In this sense, it is not a new technology derived from basic research. Rather, it is a synthesis of existing component techniques to a new level of electronic circuit function. Despite the similarity, microelectronics is having more disturbing, even disruptive, effects than did transistors on the people, structures, and relations of companies.

It is confusing the historical and proprietary specializations and blurring the relations of electronic component suppliers

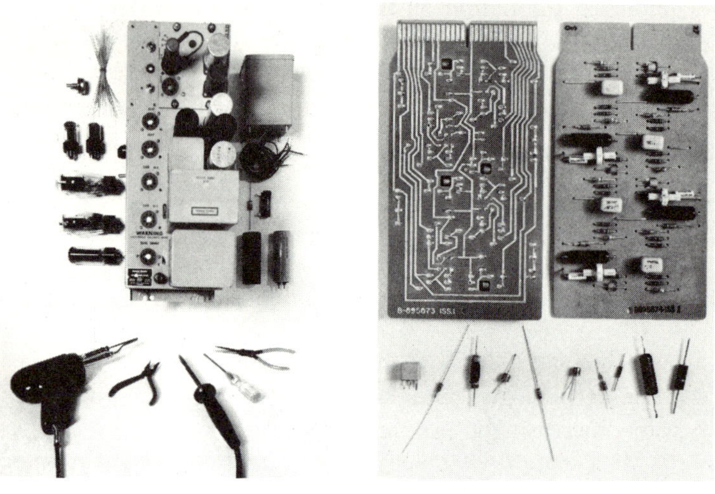
WARNING

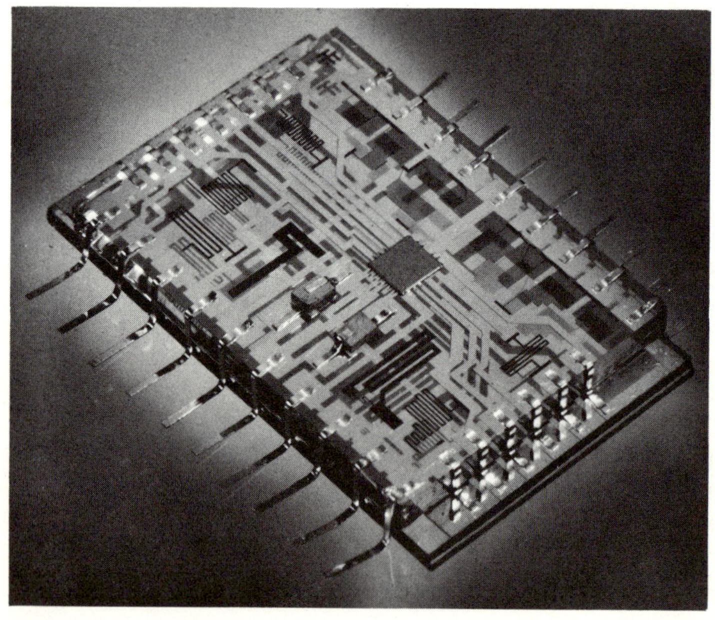

Three Steps in the Development of Electronics

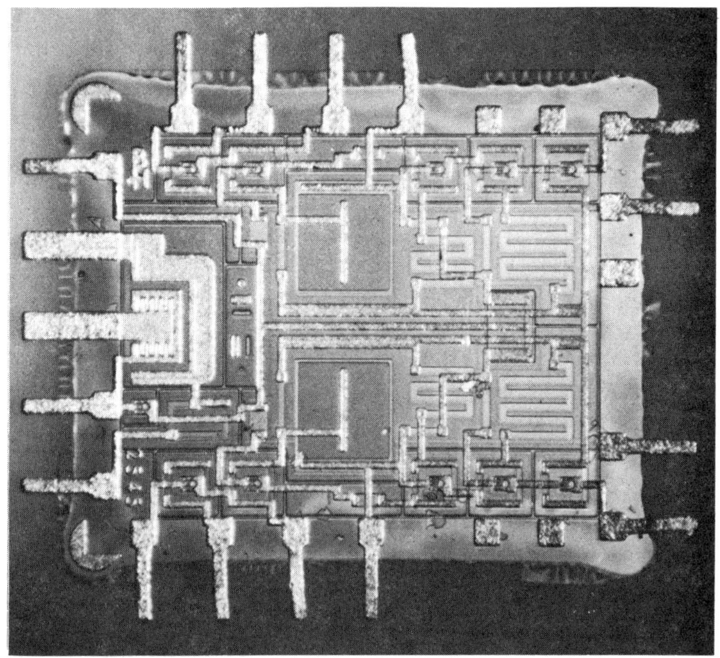

These photos signify the rapid change that has swept through the electronics industry. Top left: a feedback amplifier, circa 1950, when vacuum-tube technology was reaching its economic limit. Tubes and associated components of such amplifiers had to be made one by one and put together with hand tools. By the early sixties, the transistor had replaced the tube, as in the logic circuit for switching (top right), designed in about 1961. Such circuits had to be assembled component by component, but the components could be connected by batch-soldering on printed wiring boards. The photo at bottom shows a tantalum thin-film microcircuit, ¾ in. long, which is still under development. The high-precision resistors, capacitors, and interconnecting lands of this circuit are all made at one time; bonded in the center of the circuit (shown magnified in above photo) is a silicon microcircuit, about 1⁄16 in. long. Progress in these technologies — thin metal films and silicon — is having a major impact on the electronics industry.

and users, whether they are in single or separate corporations. Some computer and weapons system companies are integrating downward to develop and supply their own integrated circuits. Transistor and passive-component suppliers are feeling strong temptations to move upward into instruments, subsystems such as memories, and even whole systems. Even within single corporations, such as Bell Labs, component and system people are eying one another with unease.

Through these proprietary pressures for vertical integration, microelectronics is forcing a growing communications gap between component and system people. Each is diluting its former specializations; each is finding difficulties in acquiring the other's kind of specialists, and each is having trouble in getting the new kinds of products into manufacture and to market.

Specialist dilution and decoupling can waste our most precious asset — creative people with large investments in specialized education and experience. If we do not learn how to close the gap being created by proprietary forces, the innovation of microcircuit technology could be slow and costly.

During a large part of the 40-year era of the vacuum tube, research, development, and manufacturing people were loosely coupled in space, time, and organizational structure. In more recent years, we learned that the innovation process could be accelerated and sharpened in its purpose through deeper specialization and closer coupling of people. By *specialization,* excellence in depth of understanding and creativity is enhanced; by close *coupling* and good communications between the people of research, development, and manufacture, the industrial relevance of specialists' efforts is increased and the effectiveness of the process is improved.

When the transistor was invented, our understanding of the research-to-technology process was put to work to develop transistor technology as fast as possible. Despite the completely new and broader range of semiconductor science, in just 17 years transistors surpassed tubes in cost, reliability, and even in most measures of performance. I believe a great deal of this acceleration was achieved through the specialization and coupling of people *within and between* different organizations. If this is so, managers have a small share in these triumphs with the individual scientists and engineers who created the wealth of new ideas. But today we must dig more deeply into the workings of innovation in *all* its dimensions to counter the new, disruptive forces of microelectronics.

In early years, most basic research was done by university people — rightfully with little concern for its relevance to industrial technology. Their coupling to development people — people with specific industrial goals — was loose. In recent years, first in tubes and much more in transistors, much of the new science comes from relevant research in industry. It is stimulated by the industrial researcher's intimate knowledge of his company's long-term goals. This stimulation of relevant research and its rapid translation to new technology depends upon the close and timely coupling of research and development people. Microcircuit technology requires still tighter research-development coupling, because the materials and techniques are more numerous and their interactions within the microcircuit are more complex.

In like manner, tube-design and manufacturing people were loosely coupled in earlier years. Designs were developed relatively independently of the fabrication processes to be used in manufacture. But in transistor technology, it became apparent that design and process could no longer be independent and separable. For example, when we speak of a particular transistor, we specify it by its material-fabrication technique; e.g., we speak of grown-junction silicon transistors, diffused-mesa germanium transistors, and planar-diffused epitaxial silicon transistors.

The people who design components and the people who devise the processes for making them must work together in an inseparable loop of activity. Close and daily association of development-manufacturing people is essential to resolve cost-quality conflicts through design-process trade-offs. This means that innovation of technology does not involve a single act nor a random collection of unrelated activities. It is not just a flash of inventive genius, not just the discovery of new phenomena, not just the development of a new design or process — nor the creation of a new market. Rather, the process is all of these things, each above some critical threshold of excellence, acting together in an integrated way toward an overall goal. All the parts and the whole of the process stem from the individually creative yet collectively effective actions of people. It is a "people process with a purpose."

Using this viewpoint in the preceding essay, I discussed a simple model for the research-to-manufacture flow of the innovative process. Simple though it is, as shown in this sketch, it helps us understand the strengths and weaknesses of many technology-based organizations.

By viewing innovation as a people process for generating and transforming relevant research to new technology, we can see how communications barriers and bonds can be used to improve the individual creativity and overall effectiveness of people.

For microelectronics, rather than discrete components, more design-process options are possible and many more serial manufacturing process steps are required per entity. As a result, coupling between development and manufacturing people must be increased still more to obtain optimum cost effectiveness for circuits instead of components.

For component people already skilled in solid-state technology, these research-to-manufacture interactions are not new — they have already mastered them to survive. Increasing the couplings for microcircuit technology will come naturally.

On the other hand, for system companies using discrete component circuits, the flow of information has been one way — from design to manufacture, specifying only the electrical and mechanical properties of the elements and their configuration in a circuit. The feedback effect of different circuit manufacturing processes on the design of components and circuits has been relatively small. For system companies to continue their role with microcircuits, the problem of increasing design-manufacture coupling will be new and difficult. It will be compounded on top of the tough problems of understanding and control of solid-state materials and processes.

But this apparent advantage for component companies at the design-manufacturing interface could be misleading. It is true: Once having settled on a particular circuit design, component specialists do have a familiar road to travel in development, design, and manufacture. But how can component people decide *what* circuit will be optimum in *what* particular system with *what* particular set of cost-effectiveness objectives?

Early electronic systems were primarily simple extensions or joinings of smaller systems. For example, a radio receiver is a simple forward-acting system, with only a few circuit functions, such as a filter, modulator, and amplifier. Interactions between these circuits were kept to a minimum and feedback was confined to the internal workings of a single circuit. System requirements could be easily translated into terminal requirements on a few basic types of circuits.

In turn, circuits were made up of a few simple types of discrete components. Electron-tube and passive-component properties could be specified independently of their circuit and system environment. The choice of components was limited and the number of components per system was not large.

Tubes and their related components relied on classical physics for their behavior and depended mainly on the macroscopic properties of a few materials. Finally, these macroscopic (insulating, conducting, magnetic, and mechanical) properties were related to material structure mostly in empirical ways. Materials specialists depended almost entirely upon classical techniques in metallurgy.

But an effective innovation process involves not only the

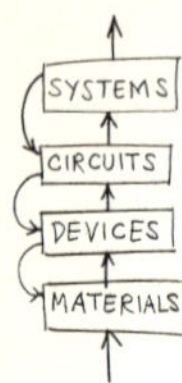

flow from broad research to specific manufacture. It also requires an information flow upward from raw materials to specific systems, with complexity increasing as the flow takes place. For tube technology, this "vertical process" was neatly divided between specialists in materials, components, circuits, and systems, as shown in this marginal sketch.

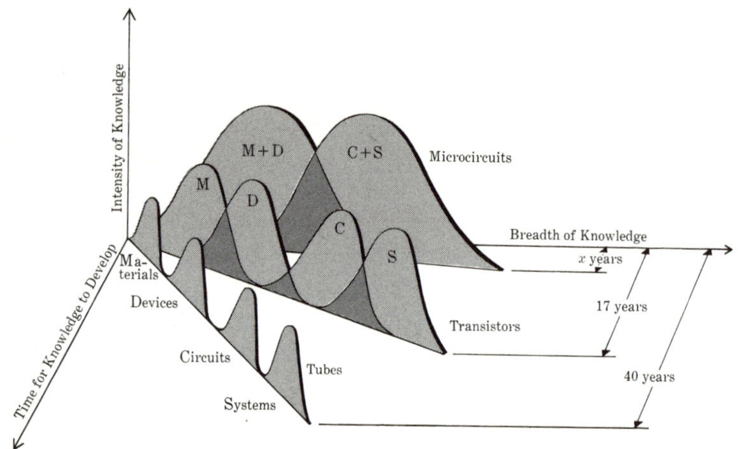

These curves suggest how knowledge intensifies and broadens as electronics progresses to microcircuits. Two things speed up the innovation process (17 yrs. for transistors vs. 40 yrs. for tubes): First, knowledge intensifies and broadens among specialists in materials, devices, circuits, and systems. Second, these specialists begin to share common knowledge as symbolized by the overlapping of the specialist areas on the curves.

The main flow of information — upward and downward — was in terms of the electrical and mechanical parameters of a few types of well-known materials, components, and circuits. At each interface, the number of options was small and new advances were infrequent. The couplings between specialist levels could be infrequent and in written form. Because of the limited number of options at each level, standardization of components and high-volume savings were possible. Completely separate companies could specialize in materials, components, circuits, or systems. This pattern of specialized knowledge activity is shown graphically in the curve labeled "Tubes." Let me show how to read these three graphs:

Intensity of knowledge specialization is represented on the vertical scale. Along the horizontal scale is represented the breadth of knowledge covered by a specialist activity. And along the third orthogonal axis is measured the time span over which the knowledge processing takes place. First, let us look at the graph called "Tubes," which shows the knowledge generation in materials, components, circuits, and systems for tube electronics. There was not much overlap understanding either in time or content of knowledge generated and used by the various specialists; and the total knowledge of any one specialist (the area under each hump) was modest by today's standards. The projection of the total knowledge activity curve onto the time scale corresponds to the 40-odd years it took for tube technology to develop.

When the transistor was invented, the materials and device people had to intensify and broaden their knowledge of the modern physics and chemistry of these new devices. Since the design and fabrication of transistors depended upon the structure of the materials at the atomic level, overlap in understanding and coupling of materials and device specialists became essential. Today, the transistor engineer must know the physical chemistry of his materials and so must the materials man understand the devices. Material-structure-technique trade-offs permeate the daily work of materials and device engineers.

Because of the large cost-reliability advantages of transistors, electronic systems have increased in complexity by orders of magnitude. Not only have the components per system increased greatly, but so also has *the number and kind of interactions* between the various system parts. Circuit and system engineers must work together to study the interactions between the various circuits in the system environment. *The interactions between circuits* in systems have become as important to the system behavior as the behavior of the separate circuits. The complexity of interactions between circuits is approaching that which exists between the elements of a circuit. The specialized knowledge activity era is labeled "Transistors." This curve shows how the total knowledge of each specialization has increased (greater area under each hump), but equally important, it shows how the overlap in knowledge and time has increased at the materials-device and circuit-system interfaces. The shorter time projection (17 years) of the transistor history graph shows that, despite the increase in specialized knowledge, more overlap in understanding and tighter coupling in time between specialists made possible a shorter interval from concept to maturity.

In transistor technology, much of the system cost is in the cost and reliability of the components. Economical design dictates holding the quantity and types of components to a minimum. Once circuit engineers learned to think in terms of transistor characteristics and circuits, the situation became not too different from that of tube-circuit design. As a result, only a small increase in the overlap in understanding and coupling in time between device and circuit people has been required. The traditional corporate separation between component and system organizations has been able to meet most of the demands of discrete transistor technology for specialist coupling.

Microelectronics, however, promises more kinds of circuit functions at greatly improved cost, performance, and reliability. System size in number of circuits, and system complexity in variety of interactions between circuits, can increase greatly and still be economically viable. System diversity can increase. The combinations of cost-effectiveness measures will vary widely for systems having different goals and environments. Such increase in the size, diversity, and complexity of system functions will call for more knowledge on the part of circuit and system engineers. At the same time, because of the increase of interactions between larger numbers of circuits, circuit work can not go on independent of system development. Circuits and their interactions must be studied, right from the start, in their system environment, and this means that the circuit engineer and the system engineer must work as one. They must be so tightly coupled that the two may well blend into a single system's discipline, as shown in the "Microcircuits" curve.

Microcircuit technology contains many more possible materials, structures, and techniques for components, and more combinations and permutations of them are possible for a selected circuit function. Greater knowledge of materials, processes, and device structures will be required for mastery and control of microcircuits — yet, at the same time, even more overlap in understanding and effort between material and device specialists will be required. They, too, will need to blend into a single more powerful discipline.

Within a microcircuit, we expect that element and intraconnection cost and failure rate will be reduced by factors of ten or more over that for a discrete element. No longer will the best choice be simply that circuit having the minimum number and types of elements with individually specified close tolerances. With the large number of options now open for materials, techniques, device structures, and circuit configuration, the "new microelectronic" component and system engineers will be forced to understand each others' technologies far more than before.

Thus microelectronics poses a dilemma. Its technology demands more knowledge and tighter coupling of its specialists in time and understanding at each interface from materials to systems. At the same time, the traditional separation of component and system organizations generates proprietary pressures that tend to force them farther apart.

Competitive systems companies hesitate to work closely with a common component supplier — especially when it is needed most — in the exploratory phases of development when trade-off decisions can be made at low cost. The systems people also see some of their former circuit-design responsibilities, sales, and profits being taken over by the component technology. They are tempted to acquire the knowledge and competence to develop, design, and manufacture their own microelectronic materials and circuits.

Conversely, the component organizations see the threat of

their business (and perhaps their specialists) disappearing into the system organizations. Having the materials-device competence and experience, they feel the capability and the temptation to move upward — not only into circuits, but into systems as well. Yet this leads them into competition with their customers. Even in "vertically integrated" companies, where no corporate or proprietary barriers exist, component and system people have similar problems among themselves.

The growth of microcircuits gives the potential for another big jump in the capability of electronic systems. To get it, we must get a corresponding improvement in our people processes. If systems — of people or things — are to become more competent, they must also become more complex. This means that systems will be made of a larger number of more specialized parts, which must be better interconnected. As our technological systems become more complex, so also must our people processes for innovating them. Increased specialization and coupling are common to both.

In striving for these advances, we must remember that the total span and depth of knowledge of any one individual is limited. No single person can cover the full range from materials to systems at sufficient depth to be creative in all areas. On the other hand, our future specialists will need deeper yet broader knowledge.

There are complementary challenges for the educator and manager as well. Education of tomorrow's electronic innovators must combine the basic sciences of formerly separate technological specializations along with an understanding of the system method of complex problem analysis. For example, tomorrow's material-component engineer must master the disciplines of physics, chemistry, and systems engineering, which are relevant to microcircuit technology. At the same time, managers must provide an environmental structure in which specialists (both old and new) can continue to increase and consolidate their knowledge in their own and neighboring fields. It is not enough to provide the opportunity for continuing education. The manager must provide equal challenges and rewards so that *all* specialists will *want* to share an overlap in understanding and will *want* to work together toward common goals.

The microcircuit history curves describe the kind of increased knowledge and coupling that our new specialists must achieve to meet the needs for innovation of microelectronics. It is a specification only of the *performance to be achieved* by people; it *does not specify the people structure to* achieve that performance.

In fact, it is impossible to specify a single people structure for all industrial goals and environments. The particular structures used by different enterprises will and *should* differ widely. Each enterprise has a different mix of goals, environment, and resources. These will change from the present mix to a future set depending upon the understanding, goals, and actions of different managers.

Whatever the goals and structure of an enterprise, whether integrated or specialized, managers must understand the innovation process. Specialization and coupling of *all* the critical functions in *both* dimensions is essential for the *total* process, whether an enterprise is a part or the whole of it. Proprietary forces should be examined with great suspicion, whether within or between companies, since they will produce decoupling, dilution, and waste of people.

Whether a management undertakes all or a part of the vertical and horizontal processes, it must achieve a level of excellence in its chosen specializations and an adequate coupling in understanding and time between all specializations, whether in or out of house.

Finally, management will have to act as judge in choosing between various materials, structures, and techniques for microcircuit technology. With so many alternate techniques and so many different possible goals, every manager will have to make selective choices. The future structure and success of the enterprise will be fashioned by these choices. In making them, the manager must consider:

The kinds of systems (or subsystems) and measures of cost, performance, and reliability specific to the market it chooses to serve.

The flexibility of the selected technology. Can it accept technical change and advance without obsolescence?

The effect a chosen technology has on the people process. Does it encourage overlap in understanding and coupling between specialists? Or does it drive them apart?

More than ever before, managers' decisions on long- and short-term goals and programs will depend upon their understanding of relevant science, its technology, and their specialized people. Only through a system approach to the people process of innovation can they achieve, or be part of, a whole structure that will survive and grow in a new era of technology. Managers who understand and are sensitive to the people problems will best determine the collective purpose and motivation, the specialization and coupling of our most precious asset — creative people.

Jack A. Morton

Science on Park Avenue

17

"What the marketer seems to be asking for is the totally new invention. Good. But what is the reaction when the new invention is produced?"

Until 1963, I had worked as a scientist — first as a senior chemist at the Oak Ridge National Laboratories, later at my corporation's research laboratory in Cleveland. But in 1964 I became involved with the world of consumer products marketing — a world that sometimes seems to have no comprehension of the function of science, but a world whose problems I have now begun to appreciate.

From my special view of these two worlds, seen now from my office on Park Avenue, perhaps I can convey some notion of the issues separating them. I have no elixir by which marketing and science can forever live in happiness. Indeed, I feel they are destined to live in conflict. Yet if the corporation to which they are mutually bound is to draw strength from their alliance, then marketing and science must work toward a mutual corporate goal. My intent with this essay is to relate my science-oriented view of the marketer's world, including some dangerous misconceptions among marketing people of the role of research and development, and vice versa. Unless these misconceptions are corrected, I fear that much of the money going toward the development of new products will be wasted and that disenchantment will prevail with regard to the value of R&D to the product-oriented enterprise.

I said that marketing and science must work toward a mutual goal . . . and, indeed, the effort is made; but when I hear the dialogue it almost seems as though two languages are spoken. Montaigne once wrote that wars are caused by the ineptitude of diplomats when trying to convey the messages of their kings. I often think of Montaigne.

We, the researcher and the marketer, cannot even agree on a definition of a *new* product, or I should say an *ideal* new product. The marketing man tells us that a product should have uniqueness, that it should satisfy existing or latent consumer needs — and in a way that is demonstrable. Ideally, the product should be protected — either by patents or know-how.

To the man in the laboratory, these are laudable objectives: What the marketer seems to be asking for is a totally new invention. Good. But what is the reaction when the new invention is produced? The marketer says: "This is not ideal. The market for this idea is questionable. Its potential cannot be measured."

So the idea the marketing man is really seeking is the modification — something better than that which already exists, something so superior, in fact, that it will monopolize the market, but for goodness' sake, not a *brand*-new idea that involves plowing new ground. Now, before we conclude that marketers are clods and that it is hopeless for technical people

"Caution toward the new idea is caused by bitter past experience."

to try to work with them, we must determine the cause of the marketers' caution. And when we do, we find that marketers must work within budgets, that the problems of market creation involve changing human habits, that product promotion is fraught with uncertainty and financial risk, and that eight new products in ten fail. But beyond this, we find that caution toward the new idea is caused by bitter past experience with groups within the laboratory who promised more than they delivered. In the business world, therefore, laboratory people live under a cloud of mistrust inspired by the ideas that they said were "sure fire" but that never made it to market because nobody wanted them, not even the samples the marketers tried to give away.

The success of scientists and engineers in producing radar and the bomb was taken as a sort of universal proof of the efficacy of post-war science to produce civilian hardware. Departments of research and development were essential to any company that expected to grow. Many with no particular link to atomic science, for example, allowed research departments to grow up in fields allied to the atom. Regardless of the businesses they were in, these firms had faith in the atom and lived in the hope that this gesture would usher them into the "atomic age."

But the score sheet for successful utilization of scientists in new-product development was dismal in comparison with their work in the war. It would be an exaggeration to say that all business executives blamed the scientists exclusively when postwar work did not yield the glories of their war-years' dreams. But it becomes ever more apparent that the bloom is off the rose. And if we are not to be blamed that others' dreams did not come true, still we bear a responsibility for allowing a misunderstanding to exist — a misunderstanding of what we *could* produce.

Although the success of wartime research was not really relevant to the problems of industrial research and product development, we scientists implied such relevance by our silence, if not by overt claims. We knew that wartime projects were aimed at specific goals, that funds were unlimited, and that the pick of allied world science was available. Seeming miracles were produced, of course, for we had these marvelous advantages plus the drive and dedication that the war inspired. We cannot liken this to the world of business competition, for the basic elements are missing. And yet we allow the miracle myth to linger on, though hampered now by disbelief. And we cause further disenchantment by failing to make management understand the nature of our laboratory problems and the probability that we will solve them. I will give an example in a moment, but first I want to stress that I am not simply talking about a problem in "communication" — though our "jargon" or inarticulateness, take your choice, is a weakness in many of us. The problem seems to me to be one of inability to convey technical facts *plus* a naive, unjustified enthusiasm for our capabilities.

Let us take a hypothetical case: a new product development in a company that has decided to diversify. What kinds of problems are likely to arise? (Some happen only once; others become habitual.) For example, suppose that we gather in executive session — a top manager, a marketer, and a research director — and agree that we wish to develop a new hair spray. The marketing man reports the need for it; the top manager asks the research director to investigate. In the laboratory, the director assigns the task to his most successful and capable group leader. He and his crew examine existing products, making chemical analyses to determine constitution and testing performance to get fixes on defects. Following the current laboratory fad, the group leader files his first biweekly progress report. It is his usual brilliant and reasonable analysis of the problem and, hence, liaison between business and the laboratory is reaffirmed. Such is the purpose of these reports, but their consequences and inferences are the occasion of sin, for after two or three reports — within six weeks — the group leader reports, to the amazement of all outside the laboratory, that he can duplicate the product of the leading competitor.

Spurred with enthusiasm, the group leader produces another report announcing this approach toward the development of a new ingredient: When formulated with standard ingredients, common to existing hair sprays, the new chemical will provide a product that is superior to any extant. The report implies that the new product will be so manifestly superior that no sensible consumer will want to use competitors' products ever again. The director of the laboratory approves this tack, though he adds a note of caution and edits the report to tone down its more rosy speculations.

Now others must dig in. A product is coming, so markets must be analyzed, sales groups must be set up, production must be scheduled. All of these must mesh with the laboratory group leader's timetable for producing the new, magic-ingredient formulation.

But the new formulation will not cooperate: It won't allow itself to be synthesized.

Several reports come out of the laboratory — all highly technical — more technical than those early reports. Management focuses on the reports and perceives delay. But there is no apprehension. Delays must be anticipated. The laboratory director interprets the latest reports to his management colleagues, the gist of his interpretation being that his group is working with a difficult problem and that delays are bound to occur.

Meanwhile, group leader and crew succeed in synthesizing an intermediate compound. Their hopes rise, but then the new compound hydrolizes in the damp air of the laboratory. Special air-conditioning equipment must be installed. Another delay.

"I have no elixir by which marketing and science can forever live in happiness. Indeed, I feel they are destined to live in conflict."

And further laboratory problems.

Because of his difficulty in producing the new material, the group leader shifts his focus slightly. His product will do the job — never worry — but the formulation will not be quite the same as forecast in the early reports. And sure enough, after titanic struggles, the new material is produced in the laboratory: Applied to a wig of human hair, it works beautifully.

The time schedule is off, but no matter, for the major problem is solved — the laboratory curiosity is now a product.

But is it? The new formulation turns out to work well, except that it will maintain stability as an emulsion for only six months: after that, it becomes unusable — at least in the aerosol formulation for which it was intended. But it *must* have a one-year shelf life — "residence times" in warehouses and among wholesalers and retailers average one year for such products. And here the laboratory group issues its final report: Because of difficulties of shelf life and stability, they recommend that the product be sold as a paste, rather than as an aerosol spray. The new paste product will still have its significant advantages over existing spray products and, in addition, it will enjoy a longer shelf life than *any* aerosol.

But, say the marketers, the product we had anticipated was to be a creamy liquid, a spray. This is what our market data called for. So with a bow to technology, they agree to make some market tests. And the tests bring gloom, and finally a conclusion, to the miracle paste. People will not buy it.

Doggedly, the laboratory group persists with its own tests and proves, at least within the group, that the paste is superior by any standard. For instance, the wives of the group think it is marvelous!

And consequently, the laboratory group is convinced that its product is a victim of bad promotion. But the marketers feel bilked because the laboratory did not deliver a product that would sell.

What happened?

Clearly, the reports did nothing to communicate the status of the project — rather, they heightened confusion. But — a vital question — did the reports actually promise successful production of a superior product? If we examine them now, we find no explicit promise, for the scientists knew they could make no such guarantee. But somehow the people outside the laboratory drew the notion from those reports, and perhaps from the enthusiasm of the laboratory director, that such a promise *had* been made.

Some will argue that the director of the laboratory was

guilty of bad management practice. I will not dispute this. Some will say the chances are slim that new products can come from developments that are prompted by evaluations of market data. I tend to agree, for I know this inhibits creativity and leads to the "me too" product.

But in the real world of product-oriented research, the people who perform the laboratory functions cannot escape the responsibility of considering the *total* corporate problem. And a similar responsibility belongs to the people in marketing and top management. Within the laboratory, this responsibility has two parts. One is the day-to-day task of product improvement. This is not the most stimulating part of laboratory work, perhaps, but it is an essential part of product development. The other is new-product development — the effort that will take the corporation into new fields. I shall not attempt to say which is more important, for each is essential, but I do say that one of the major problems in laboratory management arises when these two functions are confused with one another. My example with the hair spray is an instance of this: The group assigned to the problem was thought to be at work on a much-improved modification of an existing product, when actually it had changed course along the way and was developing a new substance.

I said earlier that the marketer wants products that the consumer can see a need for, products that really satisfy wants. Now, the hair spray example is a violation of this: The people in the laboratory were developing a product that *their* consumers, i.e., their wives and friends, approved of. What laboratory people often forget is that the *real* market is different from their microscopic market. For instance, the "clean chemical smell" admired by the scientists (and perhaps by their wives) may be obnoxious to the housewife. Further, the product characteristics that are measurably unique in the laboratory may not be so readily recognized and appreciated by the consumer, and hence will have little market value.

It is too oversimplified to suggest, as a solution to such problems, that corporations assign competent executives to oversee both marketing and the laboratory. Well-managed corporations do this as a matter of common practice. The executive, together with his managers of marketing and research, designates corporate goals and product-development areas. And, to be sure, the group communicates frequently over mutual problems. In short, they follow the procedures of good management.

"I do not suggest that the laboratory waits for top management to tell it what to do . . . but if the people in the laboratory are to meet their responsibilities, they must know the organization's needs."

But this does not delineate the responsibility of the research director. Lowell Steele has described him as "a man in the middle," as one whose position required that he establish a viable accommodation between the scientist and his organization. I agree with this description and in the context of my essay I would add that the director of research has the further responsibility of the envoy: On the one hand, he must tell the people in his laboratory what the needs of the organization are, what new markets look promising to management. He must define these needs and promising markets in ways that will enable his people to direct their technical capabilities toward the corporate objectives. I do not suggest that the laboratory waits for top management to tell it what to do, but people in the laboratory feel a deep responsibility to the organization, and in order to fulfill it they must know the needs. The laboratory director hopes his guidance will direct his scientists toward solutions of relevant problems. But sometimes their instincts lead to seemingly irrelevant areas. And here the director must make a management decision: Is this new area worthy of investigation? Will it lead us anywhere? Or does it take us too far afield? This, in brief, is his role as a manager of a laboratory. Most managers are good at this.

On the other hand, he must perform as the technical man at the business conference, and here he often fails. There is the hair spray fiasco, for example, where he neglected to describe the difficulties his men were encountering. And there are the "lectures" to his nontechnical colleagues on scientific aspects of laboratory problems, which make him seem to be talking down to them. This becomes a tremendous disadvantage to him, for he will be hampered later on when he attempts to gain the confidence of these same men as participants — and equals — at general meetings.

As the laboratory envoy, the research director is expected to relate those "over the horizon" developments that have begun to show promise. His business colleagues want to hear of these, and in reporting them the director can perform what is perhaps his most important function — as the man who explains what the laboratory is for. The marketing men will seize on these new ideas and ask that they be turned into "breakthroughs," for the marketer's vision is fixed on tomorrow, not five years from tomorrow. If the marketer cannot have the really new idea tomorrow, he will ask for something that *is* within reach — even the "new, improved" modification. But the laboratory must carry on both kinds of activity; otherwise, nothing but modifications will ever emerge, and it is the research director's responsibility to see that this is done, that his

R&D groups are not *all* fighting fires. The effective research director is the man who does more than serve as an envoy: He is a powerful influence in the world of research and in the world of marketing.

Albert C. Stewart

18 Designing a Technical Company

Since 1950, nearly 2,000 people have tried to establish technology-based companies. Only about one in eight of these new firms lasted ten years. The others have failed or undergone forced merger — or the founders have relinquished control in exchange for needed capital. To be sure, many of the survivors have prospered, but why did so many of the original 2,000 fail?

The technology-based company has received a good bit of study from academic and business people — people whose interests are grounded in economics, administration, psychology of organizations, and engineering management. In the course of these explorations, some special properties of technical enterprises have been discovered, and the poor survival rate of such firms can be explained in terms of these.

The most important of these properties is that the technology-based firm depends on change in its environment for its survival and growth. It must be responsive to change and be able to adapt itself to its chosen environment. Few traditional firms face this challenge of change in like measure; few firms, including few technology-based companies, can respond successfully to change.

In its response to change, the technology-based firm must take special pains to define its product, to design its internal structure, and to manage its creative personnel in ways that allow it to respond profitably to new opportunities in the world outside. Failure to do one or more of these things accounts for many of the failures. Thus, careful attention to the techniques that have been devised to cope with them should help the technical company to avoid these common traps.

There is no master plan. Any company is a tremendously complex social system having hundreds of interrelated variables. Even when presented with histories of failure on the one hand and successes on the other (such as Perkin-Elmer, Polaroid, Hewlett-Packard), it still is not possible to say with certainty what specific variable makes the difference between success and failure. But by pulling together the diffuse literature and adding a generous amount of practical experience, an outline of successful performance emerges.

Let us say you are a manager in a firm in the analytical instrument field. I choose analytical instruments because it is a field I happen to know something about, but I believe the generalizations we make about it can be applied to other technical fields as well.

Your business must be planned for change, for you know that a typical new product design will be obsolescent about five years after it is first marketed. Ten years from now, you will be producing things that do not exist today — probably not even in concept. The markets for most of the products you will sell ten years from now are either latent or nonexistent. This means that the product idea you have now must be only the first of a continuing series of new product ideas, new designs and methods, and new markets. You, the manager, probably originated the first new developments, but when the firm has been running for a few years, new ideas will come mainly from the staff, because you will have other things on your mind.

Thus, your firm exists in a field where its primary asset *and* its central concern must always be the effective use of its creative manpower. (As we shall see, the range of "effective use" is very wide indeed.) Also, the growth rate of the firm will largely be determined by the manager's ability to control the ratio between expansion of facilities and the rate at which creative manpower can be absorbed effectively.

Creative manpower has a peculiar property: Whereas a 5 or 10 percent improvement in the efficiency of a manufacturing process is a major accomplishment, it is possible to vary the output of creative talent by factors of tens or hundreds. How? By varying the environment in which the talent functions. If creative output is high, the firm can afford many small inefficiencies and still maintain profitable growth. But if creative output is very low, even an efficient firm will be starved for material — and its output will be correspondingly low. Clearly, high creativity must — somehow — be designed into the firm.

Let us specify our fundamental problem more precisely: We must design a limited social system — the firm itself. This system has inputs of information, money, materials, and people. Our design must include a general transfer function, which operates on these inputs. The purpose of the general transfer function is to tell us how to get creativity out of our people and how to apply their creativity to the other inputs of our system. This is a complicated job of social engineering, an area where "best practice" is what successful managers do, not one where rules are already in the handbook. Innovative people are going to operate in *some* environment, whether we design it or just let it grow. Because we know from observation that people are consistently more innovative in some environments than others, we are interested in designing an environment that actively helps us to translate latent creativity — through the research, development, and manufacturing process — into the maximum number of new and profitable products or ideas.

As we approach the design of the general transfer function, let us begin by seeing how our problem differs from a parallel problem in a more traditional business. For example, suppose that we were in the business of producing extractive products: Our big problems would be problems of raw material supply, process cost, and distribution; in other words, the thing to be transferred would *not* be anything so intangible as an "idea." But in our case, the output can be any one of many things — anything within the realm of our technical imagination. This is good, in the sense of the latitude it provides us, but it is also a danger. To control our resources and minimize the danger, we must build a suitable input-output mechanism — a three-step process.

Step One: Defining the Output

This is usually called "defining the business you're in" — a phrase so often used that it sounds like a cliche. But it still is not done often in the instruments business, and I suspect the same is true for technical firms generally. It is not done because many managers do not know how *to* define their businesses.

Almost nobody in the analytical instruments field is primarily in the business of selling hardware. Indeed, the successful firms have uniformly recognized this fact. I remember the derisive howls from the industry when Tektronix first brought to the oscilloscope market a $250 model. The industry howled because this was a "$90 market" — everybody knew you could not sell oscilloscopes costing three times that. But Tektronix had realized that an instrument traditionally used as an indicator could be used as a measuring instrument — provided you built the necessary precision and accuracy into it, provided you demonstrated the possible applications for measurement, and provided you taught people how to use it.

If you make and market analytical instruments, what you really offer is analyses, information of a particular sort. The fact that your instruments are capable of the most sophisticated analyses will not, of itself, guarantee your success. The instrument must produce the information each customer needs, on his time scale, in a form he can use — and he must be able to make all this happen reliably.

This is why the successful firm adopts some form of systems approach to its profitable marketing of information. Among other things, the firm pays attention to peripheral equipment (as well as primary equipment) and "software" (as well as hardware). In its approach to "What business are we in?" the successful firm recognizes that good technical manuals, service support, and specific application data are integral parts of the information processing systems it sells. Such operations are still looked on as necessary and costly evils in some places, but the smart managers know that these things demand the same careful attention the hardware gets. Why do you think IBM spends all that money training programmers for other people to hire?

It turns out that, when support functions are properly designed into the firm's output, they often become profitable in their own right. In the instrument business (and again, I believe this is true elsewhere), it helps to think in terms of profit per hundred analyses, or per unit time, instead of profit per sales dollar. I say this because it then seems less important whether one leases hardware or sells it, or what fraction of factory cost goes into "support" functions.

As a technical manager, you are constantly concerned with defining wanted results: products, relative importance of profits, capital gains, and growth. Because the inputs to your firm are quite general — people, money, equipment — your first step in designing a system is to list the wanted outputs as completely and specifically as you are able. Then you work backward from the desired outputs to construct a system that will produce them.

Step Two: Designing a Social Servomechanism

In place of the concepts and language of traditional management — line and staff distinction, equality of authority and responsibility, and rigid departmentalization of function — I want to introduce a more useful and congenial set of concepts that treats an organization as a "black box," an organized device with limited interfacing to the outside world. In other words, I'll treat the firm as a servomechanism, having the inputs I described as money, information, and people, outputs of information and product, and the two connected by the overall transfer function of the system, which can be specified in the usual terms: amplification, feedback, gain, delay, bandwidth, distortion, noise and bias, and natural frequencies.

A business — or any social organization, or a person — does not behave in a linear, preprogrammed fashion when it responds to a changing interface with its environment. It begins in a certain predetermined direction, then constantly samples the environment and adjusts its behavior to conform to the changing environment in a continuous process under feedback control.

Let us take one example. A central concern of your business is to take a design idea for which there seems to be a market and to translate the idea over time into a product for the market. You test the market by asking some potential users if they would buy the device. The information you get back is distorted, because the users do not fully understand your idea — or you their answers. The answers are biased, because only in special cases will your sample of users react identically with the market as a whole. Thus, the information is amplified, either positively or negatively, because the users at this stage can't possibly know the real potential of the device they haven't yet tried; they will overreact in some way, depending on personality. (You have probably had the experience of hearing one customer who *really* wanted you to build something he needed; he asked so loudly and so often that he sounded like an entire industry at your doorstep.) Finally, market information will have a delay built into it, because nobody can perceive the real market instantly. All of these errors in perceiving the real state of things can be explicitly diagrammed, approximate quantitative values assigned to them, and their effect assessed by evaluating the way they affect the firm over time.

As you proceed from the original design idea through development, you will constantly test the market; the response you get each time — poorly as it may represent the real world — will affect the way you design the actual device, the market value you think it has, and the performance you build into it. The final result will not look very much like the original idea. This describes one process within your firm, having a single first-order feedback loop through the outside world; the firm as a whole has many of these processes, with many interrelationships. If you describe them all, you have described your firm.

If you were to make a simulation model incorporating every important element of all the processes that together make your whole firm, you would be able to compare the output predicted by the model with the results you wanted. If they were vastly different, either the model would be wrong or your intuitive design of the firm would not, in fact, produce the wanted outputs, and you would have to go further and find out which situation was true, and why.

The best developed and (for our purposes) most usable description of organization design in these terms is the systems modeling technique used in industrial dynamics, developed by J. W. Forrester at MIT; much of the work that has been done in isolating important design criteria has been done by Forrester and his associates. The technique and its notation are easy to learn in principle; the results often raise questions of considerable subtlety. In use as a research tool, an industrial dynamics model of an organization is translated into a set of related equations usable by Dynamo, a computer program written especially for the purpose. The Dynamo program is then run on a computer to simulate the dynamic behavior of the model over time.

You may never simulate your firm on a computer. (Though guidance toward learning how to do this is given in the back of this volume, in my To Dig Deeper suggestions.) The models that have been constructed and published demonstrate that a model sufficiently detailed to represent the behavior of a major part of a real firm dynamically is a set of equations — corresponding to the delays, amplifications and other parameters we have described — usually representing linear and nonlinear variables in number between 150 and 1,500 and containing high-order feedback loops. The equations are only partly drawn from the financial and statistical data available in the usual management information system. The rest are quantitative statements (precise and hopefully accurate!) of such variables as our delayed and biased market information, and the interrelations between them.

You may now be asking: "If making a good model is such a complicated project, a project I may never undertake, why should I be interested in the technique at all?" The answer is that, even if you never perform an experiment, never build a model on a computer, you can still get more than half the good from the systems modeling concept from a small investment in time. I say this because I believe that once you understand the *concept*, you will think of your firm in a different way. You will not think of it as a linear, preprogrammed mechanism presided over by a vertically constructed organization chart; rather, you will see it as an interlocked set of continuous processes with inputs and feedback loops that can be, when necessary, isolated, quantified, and treated as specific design problems.

To design an organization — in the sense that I am describing it here — is a relatively new concept. The reason it has been so seldom done in the past may be the considerable complexity of the task. People are not very good at designing stable social systems: Think, for example, of the ups and downs of the economy; the high failure rate, through history, of political entities; the high failure rate of business enterprises. Further, people are not very good at predicting what will happen when they impose controls on their social systems — controls that are intended to make the systems more stable. And yet, I suspect that you are reading about the concept with fewer alien feelings than a nontechnical person might have. I say this because system design is an integral part of modern engineering; if you feel strange about the concept, it is not because of *its* novelty, but because you may not have considered its applicability to economics, government, or management.

When you design a technical system, you follow the same steps I have been describing; you state your objectives; you consider how the elements are related to one another — on through to the preparation of a model with which to test your proposed design. What I am proposing is extension of your skill in designing technical systems to the design of social systems.

To illustrate the complexity of a social system and the difficulties inherent in the intuitive design of one, let us look again at the source of our modeling technique, servo design. A sophisticated engineering servomechanism might have several hundred active elements, and up to tenth-order feedback loops of significant effect. No experienced engineer, given a range of input and corresponding output values and a wiring diagram, would attempt to make a change in its components, intending to achieve stable operation, without performing some fairly elaborate calculations. Nor would he attempt to predict the dynamic output of the system by inspection.

The business models developed in industrial dynamics rarely have fewer than 150 components — sometimes 1,500 — and rarely less than tenth-order feedback loops. The same incapacity of the system designer applies to these models as to the mathematical models of a servomechanism — predicting the output intuitively is simply beyond the capacity of the mind. Since we want to use the modeling technique to increase our design capacity, it is helpful to have some idea of what that capacity is.

When a person is considering a question, it turns out (on experiment) that the maximum number of interrelated variables he can manipulate is only about six or seven. (This echoes the comment of Ray Hyman and Barry Anderson, the psychologists who discussed problem-solving in their earlier chapter.) Hence, if there are more variables in the question, and if the question occurs repeatedly, the person will make simplifying classifications, based on his experience, to reduce the number of variables to a more manageable number. These become "rules of thumb." They work well when the situation under consideration is similar to earlier situations — but the rule of thumb is a risky guide in a changing environment.

In the hands of skilled executives, the rule of thumb based on long experience works well enough in industries that are stable; here, the environment is such that experience accumulates in sufficient quantity to make intuitive rule-making possible. It works wonderfully well if you want to make a railroad run on time. But in a technological business, a business based on change, it works less well to the extent that there is little opportunity to accumulate experience over time. The decision rules must be formulated to fit each situation; they must be tested constantly against experience, and adjusted until the desired output is obtained. There is no other way to adjust to a constantly changing world *except* by performing those operations that are implicit in the modeling technique.

For example, most companies making analytical instruments use "short-run scheduling" to allow one production line to make several different devices in serial order. The system is often installed by outside consultants, who use rather sophisticated mathematical techniques to minimize the parts inventory on hand, finished inventory on hand, and production line downtime, *based on a projected product-run size,* which you furnish after estimating the demand. In most businesses there are frequently bad estimates of future demand. Since the factory starts ordering six months before a run, a bad sales estimate must lead to too high a parts inventory (parts held over to the next run), too high a finished inventory, or too long a delivery time (customers must wait till the next run).

If, on the day you placed materials orders, you could talk to every salesman about every item, you might minimize estimating error. But if you have dealers and distributors it takes time to collect the estimates, and the delay introduces errors, which tend to propagate. For instance, a dealer who lost an order last time, because you were one piece short — his piece — and delivery therefore went from fifteen days to three months, will be very likely to overorder next time so this doesn't happen again.

Good sales managers try to adjust for error on the basis of experience by making an intuitive model of how their sales force behaves. They might just as well model this behavior explicitly and thereby derive the benefit from doing so.

The benefit is considerable. With the advent of computers, which can manipulate a very large number of interrelated variables in a short time, we can abandon the search for analytical simplification of complex systems, because the human limitations no longer hang over us. Now we can use detailed modeling, in which the dynamic interrelationship of multiple variables can be observed over time. It is only a step from observation to design — but design in terms of the *model,* which allows us to take several different approaches to the control of social systems. For example, now we can consider:

The concept of designing a system.

The concept of the firm as a dynamic social system.

The concept of feedback control of the firm.

The clear distinctions between making policy (a function of system design) and making decisions (a function of signal levels within the system).

The dominant place of people (and the communication channels between them) as parts of the system.

When we look at the firm now — taking into account the multiple variables and the dynamic interrelationships among those variables — we acquire a healthy suspicion of "rules of thumb." This is one of the major benefits of this new approach. A second benefit, even more important, is that we can now include many more variables in the decision process. The clear structuring of the decision process allows us to manipulate more variables explicitly than would be possible if we were making our decisions intuitively.

By substituting the explicit design concept for intuitive design, the manager escapes the "lowest common denominator" features of traditional management. This is particularly advantageous in the area of personnel policy, for it is now possible to design a firm that resembles the type advocated by Rensis Likert in his essay on participative management. This firm is more open, less authoritarian, less rigidly regulated socially, less constrained in its formal communication channels. And these are precisely the qualities the firm must have in order to maximize the creative output of its technical and management people.

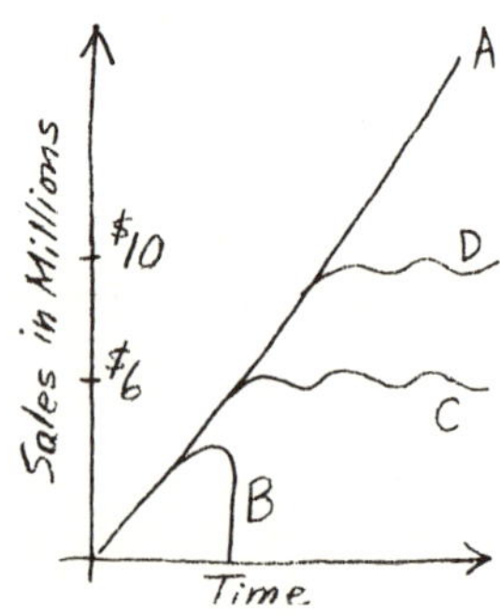

Step Three: Designing in the People

The familiar graph I have sketched here is intended to show four classical growth patterns of newly founded technical companies. We all know the names of the few companies that have grown as curve *A* — but we know, too, that they are few indeed. Many more suffer the fate of curve *B*, failing to make it through the early years. The technology-based company that follows curve *B* — either failing or going into forced merger — usually does so because its first or second product was too late to market, or because development costs were far higher than predicted or so far behind the state of the art that sales revenue could not be generated fast enough to cover current cash requirements. When such a firm fails, its failure is usually diagnosed as "financial" — but finances are more apt to be the result, not the cause.

A company that successfully passes this first test faces its next serious problem as indicated by curve *C* — a problem of stagnation — and usually the technical founder is responsible for this difficulty. Most likely, the founder is a self-confident and rather directive person. He must have such qualities in order to start a new company — and when the firm is young there is no choice but to run every aspect of it from the founder's office or hat. This works well enough until the business has grown to $5 million or $7 million in sales per year.

If a firm succeeds in growing this large, two things are usually true of it. One, it must have attracted a number of high-quality people, especially in technical positions. Two, it has grown too large for the founder to make all critical decisions on time and in an organized way. If he cannot bring himself to turn over some of the decision-making responsibility to his better men, the delay in decision-making is soon reflected in the sales curve and within the firm itself. Typically, sales begin to level off and the level hovers around $6 million. With momentum lost, the frustrations of the good employees begin to mount: They find that they must wait for decisions that they themselves feel competent to make; they find themselves affected by the seemingly arbitrary decisions of the boss — decisions on which they feel they should have been consulted. One by one, the better men are driven out of the firm, rendering the others ineffective.

A small firm cannot tolerate this for long. In technology-based companies, especially in smaller firms, people make the difference between success and failure. I think of Jack Morton's comment: What we are describing here is a "people process" and the thing we are trying to process is information. When the capabilities of people are so critical, our situation resembles that of the academic research project: Our best people are not just twice as good as our second best; they are *incomparably* better, because they are capable of high attainment and because the environment encourages them to produce at a level exponentially higher than minimum.

What kind of person is the creative technical man? From a number of studies, it appears that the best people, by and large, are more comfortable with things than with people, which accounts for their doing the kind of work they do. In psychological jargon, they are relatively weak in "interpersonal relations," which has to do with how comfortably they can deal with others in emotionally charged situations. They will tell you that they work best in a logically defined system and that they prefer to discuss matters concerning their work in rational, unemotional terms, using logic and fact to persuade and control.

But creative technical people, like other creative people, do become deeply involved emotionally with their work; when they are at their best they draw considerable emotional satisfaction from it. This very deep commitment is the source of the "abnormally" high output of some technical enterprises; clearly it is something the firm does not want to lose.

Let us go through this buildup from the beginning. The newly founded company is usually understaffed — and yet, in many cases, it is known to prosper because all the people have high commitment to the firm's goals. In this early stage, the "divergence-of-goals" problem, which Lowell Steele described, has not yet arisen, for the interests of the people and the interests of the firm are still parallel. Indeed, that is why the people have joined the young firm. When this commitment is added to the short, efficient communication channels in small organizations, the group enjoys abnormal productivity, by the traditional standards of industry.

But is such productivity really so *abnormal?* Or are we closer to the truth when we cite as *abnormal* the low productivity of such people, later on in the history of the firm, when they find themselves in routinized, functionally limited positions, where they have no control over their own performance? It is not clear to me that there is anything abnormal about the productivity of people in the exciting early days of a new firm. I have frequently been struck, when talking to men who have grown up together with an expanding firm, by the number of times they talk about the early days, "when we had time to think up new games, instead of pushing more and more paper around." One has to allow something for the rosy glow of the past when such remarks are heard, but there is more to it than this alone. Can it be that the goals of such people have moved too far away from the goals of the firm?

How do we account for the wide range of performance among people of comparable skill? Douglas McGregor discussed the question in terms of a pair of conflicting theories of management: "theory *X*" and "theory *Y*." Theory *X* is the theory of traditional management, and it starts with the assumption that people are interested in their own goals, not yours, and that they do not like to work. They must be watched, therefore, and controlled closely, chivied on with cookie and whip. Theory *Y* argues that people do like to work and, if encouraged to do so, will be inventive, enthusiastic participants in the enterprise, from whose success they will draw great personal satisfaction.

If theory *Y* is true, as McGregor, Likert, and others have asserted, then highly motivated people should not be limited and constrained, for then they will produce in a limited way as their goals shrink to conform to the job — or they will leave to find a place in which their goals and their job make a better match.

What this leads us to is some discussion of the reward structure for the firm: The kind of reward structure we establish will determine the kind of people we design into our organization. For example, if we reward people for *not* rocking boats, if we reward them for making sales pitches to management, instead of for accurate descriptions of the status of projects, if we teach them that management wants to hear what it wants to hear, then we will gather a group of people who do, indeed, respond in this way. I think, for example, of Albert Stewart's sad tale of the aerosol hair spray. An excellent case, it seems to me.

Another classic description of the results of such a reward structure is described by Edward B. Roberts, who discusses the difficulty in estimating the real time remaining to complete a design project, when the engineers have learned to bring only "good news" to management:

Uncertainty causes the responsible engineer to look for other guides (than management) to determine when he must release his drawings. Obviously, he wants to spend as long as possible on the engineering job, because the more time he spends, the more assurance he has that all bugs will be discovered and ironed out and that no difficulty will arise due to inadequate engineering. Thus, he releases his drawings as late as possible, being careful not to put himself in the position of being criticized for delaying the completion of the overall job. In other words, so long as this engineer can assure himself that *other* critical items (outside his jurisdiction) are not ready for release, he feels under no obligation to release his own portion. The result is a rather strange process in which each engineer, responsible for releasing drawings in his area, keeps half an eye on the progress being made in the neighboring area with a view to accomplishing his own release as late as possible, just so long as he is not last. The dynamics of this situation, says Roberts, are comparable to those of a horse race in which every jockey has bet heavily on himself to come in next to last.

I have suggested that the problem of designing a technical enterprise is best approached by starting with the explicit outputs you want, then systematically working backward to build a system that will produce those outputs. In technological fields, this has been accomplished in various ways, yet the resulting organizations do share certain characteristics. If your firm comes out looking like certain independent research firms, or growth companies such as Litton, or instrument companies such as Perkin-Elmer, you will probably see something like the following:

In the first place, you will discover that you need *two* management structures:

One management structure will be hierarchical and directive — shaped like an organization chart; its purpose is to control the "housekeeping" aspects of the firm: seeing that the books are kept, that everyone is paid on time, that administrative chores are routinely followed up. Some parts of your organization will live almost exclusively within this structure — accounting and plant management functions, for example.

The second management structure will be "participative," to use Likert's term. Within this structure, each group that is assigned to work on a particular activity is selected on the basis of each member's ability to contribute, his willingness to contribute (that is, his goal and the activity's goal must be the same); moreover, for each activity within the participative organization, it will be necessary to establish communication channels with all other activities that the group interacts with.

Let us give some examples of the kinds of groups I am describing here: I am thinking of design teams drawn from several engineering groups, and product management groups drawn from research, engineering, sales, and production. In fact, there is no essential difference between these groups and others that you are familiar with: For example, your organization at this moment may have a group of people, assembled as an *ad hoc* committee, to study the problem of continuing education of your technical people; you may have another group studying personnel development, or new areas of investment, or one of a host of problems. In every case, such groups act as "expert witnesses." What I am saying is that the same approach can be taken when you are dealing with the *technical* problems of the firm. This was certainly the case with the development of the transistor, as we saw in the earlier essays by Jack Morton and David Allison.

With a participative management structure, you are going to find that group activities often cross "traditional" organization lines. This is a common and accepted state of affairs in firms that function with dual structures of management. For example, in a firm such as Arthur D. Little, which functions in this way, a vice president has been known to report to a staff engineer on a specific project. Why was the engineer given the responsibility for directing the project? Because he had the particular competence to head it.

With two structures to keep track of, the managers will wind up with many more "official" communication channels than was the case before. In fact, as you well know, every organization has many unofficial channels of communication; in the organization I am describing, these unofficial channels are explicitly recognized and designed into the system. The hoped for result is that each high-performance person, having routes to official recognition through the participative structure, will continue to be an effective performer.

I must warn you that this two-structure system is going to produce many more demands for resources than was the case before. For example, you can expect the participative groups, looking at new product ideas, to come up with recommendations — some good and some bad — for ways to invest resources in development of lines you are not now in. Groups concerned with internal management will generally be oriented toward expansion of facilities and services — each of which has a price tag. Some of these will conflict. Their sum will far exceed the total resources available to you. As the manager of the enterprise, you will have to resolve the conflicts and make the decisions on resource allocations. In other words, you are no longer running an organization in which one man has all the ideas; you are no longer a one-man company. You are now in the business of making policies; you allocate resources according to the model you have created; you ensure that all groups, among them, cover the activities that you have determined *should* be covered; your responsibility is to make allocations of resources that will maximize the output you want. All of which is a complicated way of saying that you are now designing and controlling your system by policy; you are no longer wired into a large number of decision-making routines.

You will soon discover that the dual system will provide you with enough informed inputs of information so that you — the policy-maker — have rather better chances than before of making correct judgments. This will be true so long as you *don't shut off the flow of information.* Some of that news is not going to be rosy, and if you indicate your dislike of bad tidings, people will assume that you dislike their behavior in bringing such news. Soon, the information flow will stop and your model will begin the slow drift away from reality.

An open system can not only help you in building a successful firm. It can also help to create an organization that is wonderfully pleasant to work for. And this in itself is not the least important of all the possible outputs.

John H. Hoskins

19 How the U.S. Buys Research

One of the purposes of this book, as I understand it, is to bring a note of reality to the industrial research scene: to dispel some myths and to try to describe how the technical world really works. As the folk singers might say, "to tell it like it is." David Allison has asked me to wind up this series of essays by describing the behavior of the largest single purchaser of American research and development: the U.S. government itself. This seems a sensible way to come to our conclusion, and it is certainly a consistent way, for it gives us the opportunity to dispel still another myth, perhaps the biggest myth of all.

How does the U.S. government buy research and development? We know it awards industrial firms more than eight billion dollars per year for the support of research and development activities. And we know that nonprofit corporations and universities receive an additional billion and a half dollars for such work. We know, further, that there is a formal contract-award process that supposedly initiates all this funding. I have charted this formal process — and it does, indeed, look formal, even formidable.

The Formal Contract Award Process

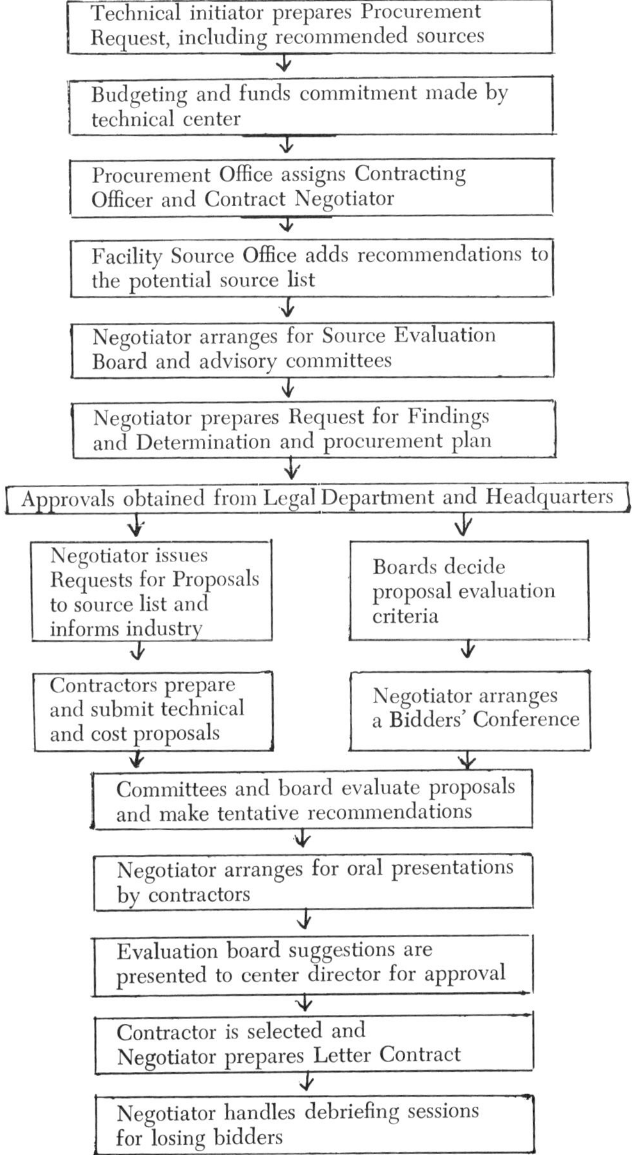

The flow of formal activities in a typical research and development contract award is supposed to work like this. The white boxes, in most instances, apply only to large contracts (exceeding $1 million).

But does the process really work the way it is described in the neat series of boxes? Is it really a filtering process, as the flow diagram suggests?

From a series of management-research inquiries I have made, under general auspices of the NASA-supported MIT Research Program on the Management of Science and Technology, I have come to the conclusion that the *real* process does not work this way at all. And this is the myth I shall talk about: that the *actual* operation of the contract-award process does not correspond to that which one would expect from the formal regulations. In fact, I will go further and say that many aspects of the formal contracting system succeed only in increasing the costs of research and development, adding time delays, and producing other damaging effects on government-sponsored research and development.

In describing how the process really works, I will compare this actual process with the formal process, so let me begin with a quick explanation of the rules and regulations that are supposed to dominate government-sponsored research and development. These rules and regulations are incorporated in the Armed Services Procurement Regulations. With only minor exceptions, the general principles of ASPR have been adopted by most agencies of the government, including NASA. In brief, the formal steps work like this: A government scientist or engineer files a Procurement Request for R&D services or equipment. In his request, he includes some specifications of what he wants to have "researched and/or developed," plus when, why, how, etc. . . . and he indicates what companies might be able to meet those specifications. Then the request goes through several stages of approval — financial, contracting, headquarters, legal — and during this process, the Facility Source Office at the government field center adds the names of other possible contractors to those recommended by the scientist or engineer who initiated the request. Next, evaluation teams are appointed, evaluation criteria are established, and Requests for Proposals are sent to companies who have made the list. Responding to these requests — or noting the announcement of the pending procurement in the Commerce Department's daily newspaper — interested contractors submit their proposals. And finally, these proposals are evaluated, award recommendations are made, and the selected firm receives a contract.

Examining the *letter* of these government procedures, one sees the principles that the formal process is intended to promote:

maximum competition;

objective evaluation — through numerical proposal evaluation by teams of evaluators;

independent, multilevel review.

But one conclusion I have reached, from over six years of research on the government R&D process, is that these sought-after principles are *not* secured. Even more general is my conclusion that many government policies for managing contracted research and development are in major conflict with the government's own objectives.

My initial studies of contract awards were made in two government agencies — one was a Defense Department center, where we examined 41 DOD awards, from $100 thousand to $8 million; the other was a nondefense agency, and here we examined 10 contracts, from $1 million to $40 million. (Another 100 awards have now been studied in a third government center and are described elsewhere.) I want to defer here to say that I was aided by Laurence B. Berger and J. Barry Sloat, graduate students in the MIT Sloan School of Management, and that we were given excellent cooperation in both government centers: We had complete access to the contracting files and personnel, and I felt throughout our extensive interviews that we were receiving *honest* — not just "cooperative" — answers. In the DOD office, we interviewed contract negotiators and some of the technical initiators and project managers. In the other agency, we interviewed the chairmen of the source evaluation boards, members of the technical evaluation committees and the business evaluation committees, and the contract negotiators, project initiators, project managers, plus others who were closely related to project award decisions.

What did we find out?

First, our file searches suggest that about 60 percent of the R&D awards were made on a sole-source basis — *without* formal competition. This finding jibed with what M. J. Peck and F. M. Scherer had observed a couple of years earlier, in their analysis of the weapons acquisition process: "In fiscal year 1959, some 53.8 percent by dollar volume of the $15.3 billion in domestic military contract awards . . . were negotiated noncompetitively with a single firm."

Now, let me point out that we excluded such noncompetitive contracts from those contract awards we studied. We did this because we wanted to know how much real competition existed in that other 40 percent — in that segment of the R&D business that got done after competitive solicitation.

What did we find out here?

We found less *actual* competition than one might have expected. For instance, of the 41 Defense Department awards, we found that the technical initiator in six of these had recommended on his procurement request that only one company be approached to undertake the R&D task. (The jargon for such a company is "desired sole source." I may lapse into the jargon now and again.) Now, acting under the procurement principle of *maximum competition,* the government agency did solicit other firms to compete for these awards: An average of seven firms got into the competition on each of these desired sole-source projects. But who got the awards? After all these proposals were evaluated — usually by small teams, dominated by the government technical initiator — five of the six awards went to the desired sole-source company anyway. (In the sixth case — the loser — this desired sole source was severely reprimanded by the technical initiator for having "insulted the agency by sending in an 'advertising brochure.'" Evidently, even a company that is "in" can sometimes get too cocky.)

In nine other cases, the government project initiator recommended two or three companies in each case. In six of these nine, the award went to one of those recommended companies — after 10 to 40 companies had been solicited in each case.

Now we have covered 15 of the 41 DOD awards. What about the other 26? In these cases, the initial recommendation lists — prepared by the project initiators — contained more than three company names, ranging up to 21 suggested sources. It was no longer as easy to assert that a narrow solicitation was desired by the project initiator. But we discovered that 22 of the 26 were *not* in alphabetical order. Did these lists, in fact, contain a preference indication? Our hypothesis claimed that, if the project initiator had his way — with formal procurement regulations absent — he would award the contract to a firm that was high on his list.

When we checked our hunch, we found that 10 of these 22 awards went to the first company listed; three went to the

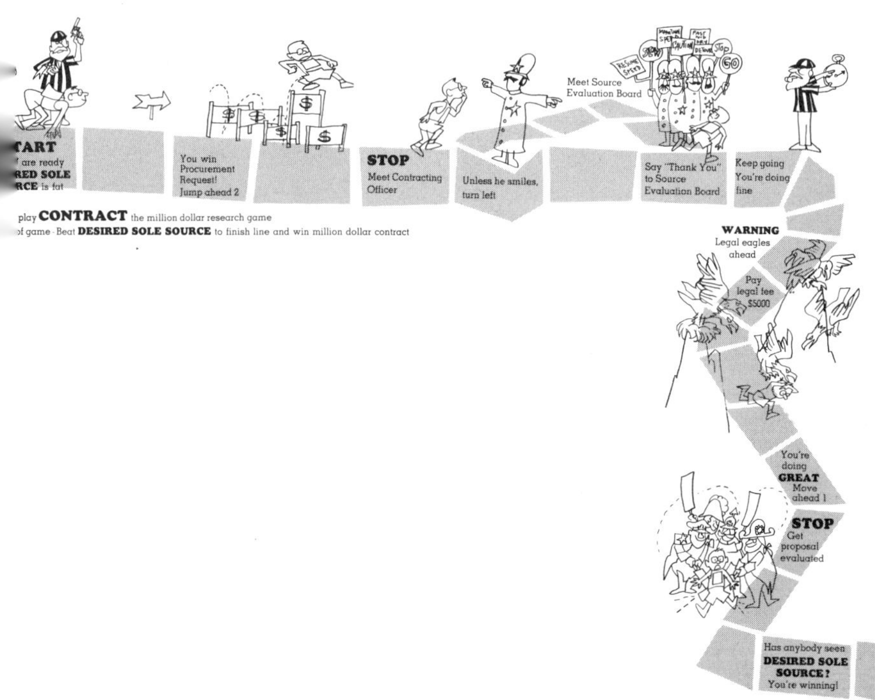

number two firm; one award went to the only recommended company that bid; five other awards went to companies the initiators had recommended. So, of the 22 awards, 19 were made to firms on the recommended lists. I do not say this is "unfair," or that initiators ought to be "more objective." I will come to the point of this in a minute, with some discussion of "objective evaluation," but I want to point out some additional facts before winding up this discussion: Over twice as many companies were solicited, in all, as had been recommended by the technical initiators; in three cases, the company that appeared at the top of the list did not even bid on the job; in only three cases did the award go to a firm that had not been recommended by the project's initiator.

But what about the four lists that *were* made alphabetically? Again, there were about twice as many companies solicited as were recommended by the initiators, but in *all four* cases the award went to a firm that had been recommended by the initiator. In two of these four cases, we found from interviews that the initiator had been strongly biased in favor of the ultimate winner, and in a third case the contracting officer said the winner could have been made the sole source. Thus, it is possible that alphabetical source lists were established merely to give the appearance of an impartial recommendation.

I must caution that these data were drawn from R&D contracts in the range of $100 thousand to several million dollars and, further, that all of these 41 awards were made by the

same agency in the Defense Department. Hence, it is possible that less "preselection" may take place in other agencies and with larger contracts, though the evidence of Peck and Scherer shows a very high percentage of *formal* sole-source awards on large-weapons systems.

But with R&D contracts of less than $100 thousand, it is likely that even *more* preselection occurs. I believe this because fewer formal reviews are required, and decision-making authority exists at a lower level of the governmental organization.

The second key principle in the procurement regulations is that award decisions should be based on objective evaluations of the proposals that are submitted. Following the formal procedure, one seeks such objectivity by getting quantitative assessments from several competent and unbiased evaluators. But our research results shows that many aspects of this principle are denied.

I said earlier that the project initiator is often biased in favor of one or two companies — firms with whom he has had contact and experience. Why do I say this? The project initiator is always a technical man. And any experienced person in the position of project initiator will have had some previous encounters with some of the potential bidders. It is hard to imagine that the initiator's judgment will be unaffected by such experiences. Because he is almost always a member of the project's evaluation team, the initiator can exercise his "predispositions." Indeed, one whose judgment *is* unaffected

is probably not suitable to handle evaluation responsibilities. Hence, I would say two things: Not only is lack of some bias *impossible* of attainment; it is probably even undesirable.

Only the naive nontechnical man can believe that technical performance and — more difficult — technical proposals can be "objectively" evaluated. In all areas worthy of the "R&D" designation, only subjective evaluation is possible. Even technical "facts" are subject to dispute by competent evaluators. Certainly, technical "opinions" on yet unproven research and development projects can be debated still more. We can expect honest appraisals by competent men, but these must reflect their experiences, judgments, technical prejudices, and other factors of a subjective nature. After-the-fact contract performance evaluations in the contracts investigated often showed that award decisions had been based on incorrect technical evaluations. One project initiator admitted that he had been "sold a bill of goods by Company X." Another project, awarded to Company Y because of its unique technical approach, was canceled when the approach proved unusable. At best we can expect competent subjectivity, not some mystical objectivity.

What about using numerical point scores in evaluating proposals? In nine of the ten nondefense R& D awards of over \$1 million, the source evaluation board members revealed that general discussion, leading to general agreement on the recommended source, *preceded* final assignment of numerical point scores. In other words, the numbers are not the meas-

ures that produce the award; rather they are after-the-fact representations of general agreements. They are justifications for decisions, rather than the causes.

I challenge other aspects of the objectivity principle. For example, company evaluation, not proposal evaluation, is the paramount consideration. Furthermore, two or three key people, at most, dominate the award decisions — not the 20 to 100 who staff all evaluation committees. One might even question the relative competence of government engineers and scientists to evaluate large system proposals, where the main criteria often relate to organization and management of a several thousand man, multicompany, contractor team.

The final premise of the formal government R&D contracting system is that several levels of review and approval ensure effective and honest award decisions. This principle also seems more an illusion than a reality.

In contracts of less than $1 million, formal source evaluation boards are seldom established. An award recommendation is made by the technical evaluation group, consisting usually of the technical initiator, his supervisor, and perhaps another scientist/engineer in the same group. Their recommendation goes to the contracting officer, a member of the agency's procurement organization who rarely has a technical background or technical review capabilities. He looks at the budget, the relative bids, and perhaps at some recent agency experiences with the recommended contractor. If he is concerned about large apparent cost differences in the bids, he might ask the technical evaluators if a lower bidder would be acceptable. But so long as budgeted funds cover the recommended bidder, the technical evaluators can have their way.

For the larger contracts, technical and business evaluation committees report to a source evaluation board, which recommends action to the center director, who (in very large awards) forwards his recommendation to Washington for a formal decision. But do these multiple review levels mean anything?

Our interviews indicate that decision-making is done — effectively, if not formally — in the technical evaluation committee or the source evaluation board. In some cases, the technical initiator "stacks" the board with people predisposed to accepting the initiator's recommendations. Here, the real power resides with the technical initiator, acting thrugh his technical evaluation committee; higher level groups in effect "rubber stamp" his recommendations.

In other cases, we found that the source evaluation board was doing the decision-making, with the supposedly independent committees under the board continuously being called upon to revise their reports until the senior group agreed with it. As one procurement official who had chaired such a business evaluation committee explained, "They tell me that, when I'm on an evaluation team, I'm supposed to be Joe Independent. What they forget is that, for the rest of the year I'm Joe Subordinate, and if my boss says 'Change it,' he's the boss."

In only one case of the ten large nondefense awards was the source evaluation board's recommendation reversed at the Washington level. The situation was one of a board split between two competitors; the board's disagreement resulted from different interpretations of the job requirements. Headquarters had to resolve this dispute, but in no other case did headquarters' interference or review appear to be important. It is possible that a sample of even larger contracts might provide more contrary evidences.

The real award process is one involving long-term, person-to-person contacts between technical people in government and industry. They build up common experiences, attitudes, aspirations, confidences. And ideas are generated in this interchange. These are the ideas that later become government-sponsored R&D projects. When he is convinced that an idea has solid merit, the government scientist/engineer initiates a Procurement Request. He often feels, naturally, that the work should be carried out by the people in those capabilities he has faith. Acting in what he believes to be the nation's best interest, he tries to secure "his" contractor. (He usually succeeds.) If he is confident of his judgment, he thwarts attempts to saddle his project with another contractor. Only when he regards several companies as being highly qualified does real competition prevail.

I do not mean that all proposals are not carefully studied for their technical content. They are. But the evaluators also consider the companies: their experience, their key people who will staff the job, and so on. What the evaluators must really look at are the contractors as potential sources of work, not just at their glossy brochures. And award decisions are reached after careful consideration, discussion, and argument have resolved the issues. Yes, these resolutions are quantified in numerical evaluation forms — but, as I say, the numbers *follow*, not precede, the point of award decision. Occasionally, a unique technical proposal or a large cost difference in the bids upsets the expected results. But far more often, the predetermined decision proceeds unmolested through the several formal stamps of approval, pausing for review only when a major unsettled dispute rears its head.

Does this process lend itself to political pressure? We looked for this, but found *no* instance of it. And I doubt that our interview subjects, who were indeed candid on other touchy issues, tried to hide information from us on this question. It may be that, because the awards we examined were fairly small — mostly less than $1 million — they did not merit such pressure. But Peck and Scherer's study, of bigger contracts, came to pretty much the same conclusion: "Political considerations have not played a really major role in the choice of contractors for advanced weapons programs." They found a lot of political "activity," but concluded that the effort did not have a corresponding effect. They said that much of the activity is "ritualistic" and that political pressures tend to cancel out one another.

In only one of the large ($\geq$ $1 million) awards did we encounter any political activity, and it was not really "pressure": Two Congressmen indicated their interest in the contract; one asked that two competing companies from his district be considered, in light of high unemployment there; the other Congressman, extolling the merits of small business, gained an audience with the center director for the president of a small company. The agency responded by tightening up internal documentation on all phases of the competition "in case of any trouble." Two of the three "congressional" companies would have been disqualified at an early stage of the evaluation, but this was not done, so as not to insult the Congressmen. (Both companies were ranked near the bottom in the final evaluation.) The company that did receive the contract was not one that a Congressman had recommended. When the award was announced, the Congressmen were notified simultaneously with notification to the bidders. Thus our only incident of political activity appears as appropriate actions by Congressmen in behalf of their beliefs and constituents; we note the somewhat ritualistic flavor, the partial canceling effect, and the ultimate lack of influence on the award decision itself. More direct studies are under way to clarify the award situation in those larger contracts.

Let us now examine both the costs and benefits of the present system of R&D contracting: We now know that the real process underlying R&D awards is informal and highly selective. But the pressure of the procurement regulations and fear of the General Accounting Office and Congressional inquiry has forced us to take the formal solicitation approach to contractor selection. This approach is bad on two counts: It is costly and it is superficial. Let me substantiate my argument, first, by citing some facts that others have uncovered.

In a study of proposal efforts on small R&D tasks — from $30,000 to $150,000 — Thomas Allen of MIT gathered data on 14 contract awards. He found that the total cost of all company-proposal efforts in each competition ranged from 3 percent to 150 percent of the direct cost of the contract awarded. The total proposal effort in those 14 competitions was not a function of the size of the job, but was related, rather, to the number of bidders: the more bidders, the higher the total proposal preparation cost.

Another study tells of 36 firms competing for a single large R&D contract: Each of the 36 bidders spent an estimated 3,000 to 4,000 engineering man-hours on its proposal. Add these together and you have 45 to 60 man-*years* of engineering effort. And these figures represent only the effort expended at the *prime* level. Then you must add the technical manpower expended by various tiers of subcontractors, who must generate technical data for the prim contractors, the bidders themselves.

And government people must also expend great effort, what with extensive source solicitation and the ritual of "going through the motions" on the proposals submitted. The key people on technical staffs often serve for weeks and months on the evaluation teams, distracted from project work in which they might otherwise be engaged.

The direct financial cost of this process is high, and government bears the entire cost of the evaluation and most of the cost (through allowed contract overheads) of the company proposal preparation. More important than this direct cost is the resulting reallocation of people: The best technical personnel, in industry and government, are pulled off into proposal preparation and evaluation. This detracts immeasurably from progress and effectiveness of the ongoing projects. It establishes a situation in which the best industry people propose and try to sell the jobs, while the mediocre are left to do the work. Furthermore, significant time delays are added to the contracting process by the necessity of the 3 to 12 months needed to prepare proposal requests, to solicit proposals, to evaluate them, and, finally, to make the awards.

These factors do not contribute to the ultimate success of R&D projects, yet current government pressures are pushing for an expansion of formal competition, and for an increase in the number of companies solicited for each project. This tendency will increase still further the direct and indirect costs of R&D procurement to both government and industry, and will draw even more engineers away from contract work toward proposal efforts.

How effective is the current regulatory system? Though hard to measure, it scores positively on several counts: For instance, one index of effectiveness might be the number of "surprises" resulting from the enforced broad solicitations. We saw earlier that seven of the 41 DOD contracts were awarded to companies not on the initiator's recommended list; five others went to firms that had been recommended, but not necessarily preferred, by the initiator. Now, some of these 12 proposals were more outstanding than those submitted by preferred firms. These were the surprises. And this is one benefit of the current system: Broad solicitation occasionally uncovers someone with a better approach to the problem solution.

But note that at least 4 of these 12 awards were made because of apparent cost differences in the proposals, not because of better technical approaches. And in two of these four cases, it turned out that the assumed cost benefits had been grossly misleading, and the government project monitors regretted the awards. Hence, we can say that cost savings do result in *some* instances, due to the present system of widespread solicitation. But this benefit cannot be evaluated with precision.

A final assumed benefit of the existing regulations is the built-in system of checks to ensure honesty in contractor selection. This is intended to prevent mishandling of government funds for research and development. I will speak to this point in a minute.

Do these benefits justify the burdens? I think not. Particularly with awards of under $1 million, I believe the costs of the regulations far outweigh the rewards. The misallocation of talent and the added time delays are more disturbing here than are the large, direct proposal costs. Further, the internal checking provided at present is both ineffective (as shown by the high degree of preselection) and unnecessary. I am convinced, from our interviews with project initiators and contracting officials, that these people are highly motivated "to get the most for Uncle Sam." The honesty of these government scientists and engineers is unquestionably high; their ethical standards have not been corrupted by the intentional-low-bidding game that current procurement regulations encourage industry to play. It seems foolhardy to continue buying insurance when the premium is more costly than the value of what is being insured.

Precedents already exist that indicate a path for improving the R&D award system. For example, the Atomic Energy Commission has visited company plants when determining the award of a contract. The Army also has relied heavily on such visit-interviews in evaluating contractors for its Sprint program. Indeed, DOD regulations permit the use of oral proposals as a means for source selection. Other informal or less formal means of industry solicitation and evaluation are possible and can be used by technical evaluators.

For R&D awards under $1 million, technical initiators should have official authority and *real* backing to solicit industry informally and to select sources without the requirement of openly solicited written proposals. The initiator, if he wishes, should be permitted to use any mix of methods, including oral proposals, plant visits, restricted solicitation, as well as the broad solicitation approach currently used. In turn, the initiator and the other evaluators should be required, after the fact, to write up and justify their method of evaluation, including a written assessment of the companies considered. Numerical evaluation should not be required, though it could be used if desired by the evaluation team. I make these suggestions in order to encourage more flexibility in source selection methods and to place greater trust and responsibility in the government evaluators. I believe these changes are in the direction in which the entire government contracting system should move.

At all levels of R&D contracting (≥ and < $1 million), a criterion for the source selection approach used should be that the "costs to procure" should relate to the costs of the procurement itself, and to the added value believed achievable by the selection method. For example, we now ask as many as 100 companies to submit proposals on R&D contracts valued at under $100,000. This practice is foolish and should be abandoned. Courage is needed in the DOD, NASA, and other agencies to resist the current trend to write cost-savings reports based on the number of contracts written "competitively," instead of as sole-source awards. Our evidence indicates that total costs may well be *increased*, rather than reduced, by such a forced changeover to superficial competition. Instead of attempting cost reduction in research and development, far more gain is possible by striving for effectiveness increases, and these increases are only possible by ridding ourselves of the constraints of the current formal regulations.

Edward B. Roberts

The Authors

David Allison was one of the founding editors of *International Science and Technology* in 1961. Technical management is his field of editorial specialization. He is a graduate of Rensselaer Polytechnic Institute, in industrial management, and for the past twenty years he has written extensively in the field. He resigned as senior editor of *International Science and Technology* in 1967 and is now a free-lance editorial consultant and writer. He lives in New York City.

Carl Barnes has been both an inventor and technical manager. He holds more than 50 patents in the field of chemistry, and for several years was corporate vice president for research for the 3M Company. Today he is president of Barnes Research Associates, of New York and New Canaan, Connecticut, consultants in re-research management, and president of ALRAC Corporation of Stamford, manufacturers of organic chemicals. He received his PhD at Harvard in 1935. His many extracurricular activities include the chairmanship of the National Academy of Sciences Committee to Study Creativity.

Jack E. Goldman is group vice president for research and development at the Xerox Corporation, a post he assumed in late 1968. From 1962, he had been the director of the Scientific Laboratory of the Ford Motor Company. He joined Ford in 1955 to head the Laboratory's physics department; he became associate director in 1959. His prior experience includes five years on the faculty of Carnegie Institute of Technology, where he also directed the Laboratory for Magnetics Research; staff member of Westinghouse Research Laboratories; and Visiting Edwin Webster Professor at MIT. Goldman is a graduate of Yeshiva University; his graduate degrees are in physics from the University of Pennsylvania. His government and professional activities include membership on the Commerce Technical Advisory Panel, the Army Scientific Advisory Panel, chairman of the advisory group on solid state science to the Air Force, the National Academy of Sciences panel on Applied Research and Technological Growth, and the Governing Board of the American Institute of Physics.

John H. Hoskins describes himself as a generalist. He has studied some physics, but he does not think of himself as a scientist, and he has done a good deal of work in technical management. Currently, Hoskins is director of grant and contract administration at Yale, which happens also to be his *alma mater*. He is an ardent observer of the worlds of science and management and a frequent commentator on the events that take place when the two worlds come together.

Ray Hyman and Barry Anderson are experimental psychologists at the University of Oregon, in Eugene. Ray Hyman received his PhD at Johns Hopkins University in 1953 and, for most of the years since, has been a member of the facutly at Oregon, where he is a professor in the Department of Psychology. In 1967-68, he was on sabbatical leave as a Fulbright lecturer at the University of Bologna. Barry Anderson also received his PhD at Johns Hopkins. He joined the Oregon faculty in 1963, where he teaches psychology and does research in the areas of learning and cognitive processes.

Lawrence S. Kubie is an eminent psychiatrist, neurologist, psychoanalyst, and writer. He has taught at Columbia, Yale, the University of Maryland, and Johns Hopkins, and has been president of such professional societies as the New York Psychoanalytic Society and the American Psychosomatic Society. His interest in creativity traces back to New York in the thirties, when he was engaged in the practice of clinical neurology, psychiatry, and psychoanalysis. Many patients came to him from creative fields — science, painting, writing, music. Out of these experiences he gained an unusual opportunity to examine the relationship between the neurotic and educational processes, and between the neurotic and creative processes. He lives today in what he describes as a converted wagonshed, near Sparks, Maryland.

Rensis Likert is director of the Institute for Social Research, of the University of Michigan, Ann Arbor. The Institute is one of the largest organizations studying organization management and development, economic behavior, political behavior, motivation, and communication. The management system he proposes is based on several hundred studies of leadership, management, and organizational performance conducted by the ISR over the past 20 years, employing quantitative methods of investigation. He received his

PhD at Columbia in 1932. He has served as president or director of many professional societies, including the Society for the Psychological Study of Social Issues, the American Psychological Association, and the International Association of Applied Psychology. At Michigan, in addition to heading the Institute for Social Research, he is a professor of both psychology and sociology.

Jack A. Morton is head of components research and engineering at the Bell Telephone Laboratories, Murray Hill, New Jersey. He was elected vice president of all device development nine years ago. After the University of Michigan (MS, 1936), he joined Bell Labs, and during his early years there he did research on microwave tubes and studied physics part-time at Columbia. When the transistor was ready for applied research, in 1948, he was put in charge of all semiconductor development work. In 1955, he was made director of device development, which gave him responsibility for both the fundamental development and development for manufacture of electron tubes, solid-state devices, and electromechanical and passive devices.

Donald Pelz and Frank Andrews are social psychologists at the University of Michigan's Survey Research Center. Pelz has investigated the scientific performance of technical people for many years, beginning with a study at the National Institutes of Health, in 1952, the year in which he received his PhD. For the past decade he has studied technical performance in industry, government, and universities. Andrews became interested in studies of technical performance when he was a graduate student at Michigan in the early sixties, and he and Pelz first collaborated during that period. He received his PhD in 1962.

Edward B. Roberts is associate professor of management at the Massachusetts Institute of Technology's Sloan School of Management, as well as associate director of the MIT Research Program on the Management of Science and Technology. He holds four degrees from MIT: two in electrical engineering, one in industrial management, and a PhD in economics. He is also president of Pugh-Roberts Associates and a management consultant to various industrial and governmental organizations, including the Technical Advisory Board of the U.S. Department of Commerce and the Scientific Advisory Board of the U.S. Air Force.

Donald A. Schon is a founder of the Cambridge-based Organization for Social and Technological Innovation, a nonprofit enterprise established in 1966. Prior to OSTI, he had been director of the Office of Technical Services, U.S. Department of Commerce, where he was responsible for the joint industry-government program on obstacles to technical innovation. For six years prior to OTS, he had directed the new-product development group at Arthur D. Little. He received his PhD at Harvard in 1955 and has taught at the University of Kansas and the University of California.

Lowell W. Steele is a student of industrial research, and his campus is that vast laboratory of social science and research management, the General Electric Company. In the early fifties, he joined General Electric's Research Laboratory, where he has served as supervisor of salary practices and has conducted studies in research environment. Recently his attention has been focused on more general problems of the management of research and development, particularly from the viewpoint of the businessman who must support the work and attempt to use its results. His essays are derived from conversations with hundreds of scientists and engineers throughout his organization. He has degrees from Dartmouth, the Harvard Business School, and MIT, where he received his PhD in 1953.

Albert C. Stewart is market development manager in the new polymers department of Union Carbide's Chemicals and Plastics Development Division. But, in a way, the lengthy title is somewhat misleading, for most of his professional life has been spent in chemistry research, including six years as a senior chemist at Oak Ridge and six years at Union Carbide's laboratories in Cleveland. The fact that he has seen science from both worlds — the lab bench and the marketplace — gives him the special opportunity to describe the interaction that takes place when these worlds come together. He received his first degrees from the University of Chicago (MS in 1948) and his PhD from St. Louis University in 1951. His technical specialties include physical-inorganic chemistry and radiation chemistry. He has been an advisor to many organizations, including the Ford Foundation, the Agency for International Development, and NASA, as well as an officer of the Children's Aid Society, the Urban League, and the Center for Urban Education.

To Dig Deeper

To find more information on the subjects discussed in these essays, we suggest you consult the references cited below. Certain of these references are mentioned in more than one chapter, where the reference has relevance to more than one essay subject; we do this, at the expense of repetition, so that each chapter will have as complete a set of references as possible.

Part I

Introduction

On the "explosive growth of science," we suggest you consider a pair of books. One is *The Sociology of Science*, a collection of fundamental papers by such men as Bernard Barber, Lawrence Kubie, and Derek Price; edited by B. Barber and W. Hirsch; Free Press, a division of Macmillan, 1962, $9.00. The other is *Little Science, Big Science*, by D. J. Price; Columbia University Press, 1963, $4.50. On science and its impact on the federal establishment: *Science in the Federal Government*, a history of policies and activities to 1940, by A. Hunter Dupree; Harper Torchbooks, 1957, $2.75. On science and its impact on science: *The Structure of Scientific Revolutions*, by Thomas Kuhn; University of Chicago Press, 1962, $4.00. On science and economic development: *Regional Effects of Government Procurement and Related Policies*, a Commerce Department document, 1967, from U.S. Government Printing Office, Washington, D.C. 20402, $0.40.

1. The Industrial Scientist

Sir John Cockroft is editor of *The Organization of Research Establishments*; Cambridge University Press, 1966, $11.50. This collection of essays by 15 directors of research — most of them British — is worth noting for the contrast it provides to the American scene; included is an excellent essay by James Fisk, who writes about the organization he heads, the Bell Telephone Laboratories. See, also, a pair of older volumes: Simon Marcson's *The Scientist in American Industry*; Harper, 1960, $3.00; an illuminating report on the environment of industrial research. And Anne Roe's *Making of a Scientist;* Apollo, 1954, $1.75; Miss Roe's studies are psychological studies of technical people. Four reprints from *International Science and Technology* may provide additional background: "Educating the Engineer," June 1963, "The Graduate Student," January 1966, "The University & Regional Prosperity," April 1965, and "Tomorrow's Engineer," December 1967; available at *International Science and Technology*, 205 East 42 Street, New York City 10017; $1 per copy.

2. Creativity

A sample of what is being talked about or being investigated under the label "creativity" is *Scientific Creativity: Its Recognition and Development*, edited by C. W. Taylor and F. Barron; Wiley, 1963, $8.95. The volume contains selected papers from the proceedings of three conferences held at the University of Utah in the late fifties. A less technical collection, *A Source Book for Creative Thinking*, edited by S. J. Parnes and H. F. Harding, will tell you who is doing creativity research; Scribner, 1962, $4.50; this book also contains a section on educational programs under way in the field of creativity.

For a view of psychologists' opinions on the subject, see *Contemporary Approaches to Creative Thinking*, by H. E. Gruber, M. Wertheimer, and Terrell; Prentice-Hall, 1962, $6.00. The contributors, all well-known psychologists, are not closely associated with research on creativity. For an understanding of the psychology of imagination, see *Imagination and Thinking*, by P. McKellar; Basic Books, 1957, $4.25; McKellar says all creative imagination must stem from previous perceptual experience. The book is especially valuable because it includes material that is typically overlooked in psychological treatments of creativity.

But all work in this field is not done under the label "creativity research." If you are investigating research currently under way, you should also be aware of the animal work of Harry Harlow, at Wisconsin, and D. O. Hebb, at McGill, especially as it relates to the role of early experience on subsequent problem-solving ability; Jerome Bruner's work, at Harvard, on how people go about attaining concepts; the work of Charles Osgood, at Illinois; Fritz Heider, at Kansas State; and Leon Festinger, at Stanford, on mental models: how attitudes are organized and how the mind reacts when confronted with new information.

3. Blocks to Creativity

If you wish to read further on Lawrence Kubie's own work in the field, we suggest you begin with his book *Neurotic Distortion of the Creative Process*; Noonday 213, 1961, $2.00. Here you will find lucid discussion of the psychodynamics of neurosis and creativity, the interactions between the creative and neurotogenic processes, and suggestions pertaining to education for preconscious freedom. On creativity in technical fields: The Spring 1962 *Daedalus* published Kubie's paper "The Fostering of Creative Scientific Productivity"; write to American Academy of Arts and Sciences, 280 Newton Street, Boston 02146; $2.00. Also: "Neurotogenic Factors in Engineering Education and in Creative Productivity," published in *Current Developments in Engineering Education*; transactions of 1961 Boulder conference; write National Science Foundation, Washington. Also: *Research in Protecting Preconscious Functions in Education*, a monograph published by the Association for Supervision and Curriculum Development, 1201 16th Street, N.W., Washington, D.C. For more general reading: *The Psychopathology of Everyday Life*, by S. Freud; Macmillan, $4.00; *Creative and Mental Growth*, by V. Lowenfeld and W. L. Brittain; Macmillan, $7.00; *Practical and Theoretical Aspects of Psychoanalysis*, by Kubie; International University Press, $4.00; also in paperback: Praeger, $2.00. On various aspects of the dropout problem, see *The Journal of Nervous and Mental Disease*, Vol. 141 (No. 4), January 1966, pp. 395-402. On the utilization of preconscious functions in education, see *Behavioral Science Frontiers in Education*, edited by E. M. Bower and W. G. Hollister; Wiley, 1967; with special attention to Chapter 4.

4. Freedom in Research

5. Diversity in Research

The best starting point for more background relating to both these essays is the December 1962 issue of *The American Behavioral Scientist*. Several current investigations are reported there in "Science, Scientists and Society," including a review of 88 studies by Gordon and Folger. For a broader perspective, see *The Sociology of Science*, edited by B. Barber and W. Hirsch; Free Press, 1962, $9.00. Also *Scientists in Industry: Conflict and Accommodation*, by W. Kornhauser; University of California Press, 1962, $6.00; Kornhauser treats the autonomy versus coordination dilemma and cites several opinion surveys of scientists. "Analysis Memo #16," mentioned in the first essay, is available from the Institute for Social Research, University of Michigan; ask for Memo #16, February 1962, publication 1900, $0.75. Two books by Rensis Likert relate to both these essays: *New Patterns of Management*; McGraw-Hill, 1961, $6.95; and *Human Organization: Its Management and Value*; McGraw-Hill, 1967, $7.95. Also: the authors' own work: *Scientists in Organizations,* by D. Pelz and F. Andrews; Wiley, 1966, $10.00. Further details about the effects of spending full versus part time on research appear in Andrews' article in the September 1964 *Administrative Science Quarterly.* Also: Pelz's "Creative Tensions in the Research and Development Climate" was published in the July 14, 1967, issue of *Science.* And the authors' "Autonomy, Coordination, and Stimulation, in Relation to Scientific Achievement" appeared in the March 1966 *Behavioral Science.* For further data, not available elsewhere, see "Supervisory Practices and Innovation in Scientific Teams," by F. Andrews and G. Farris, in the Winter 1967 issue of *Personnel Psychology.*

6. Solving Problems

Many books give advice on how to solve problems. Psychologists cannot prove with experimental data that all such advice (or even most of it) is going to be helpful, but almost any systematic procedure may interrupt habitual modes of attack and thereby pave the way for finding better problem-solving alternatives. Some general books worth considering: *The Art of Clear Thinking,* by R. Flesch; Collier, 1963, $1.00; many of the author's hints overlap with Hyman's and Anderson's precepts. *The Art of Problem-Solving,* by E. Hodnett; Harper, 1955, $4.00; this is a thorough classification of the various points at which the problem-solving process can be improved. *How to Solve It,* by G. Polya; Doubleday Anchor, 1957, $1.00; a classic, based on attempts to nudge mathematics students into creatively finding proofs for difficult problems. *Notes on the Synthesis of Form,* by C. W. J. Alexander; an interesting book, and one that is apt to be overlooked by nonarchitects; Harvard University Press, 1964, $7.00; if carefully applied, Alexander's approach will tend to prevent premature assimilation of a problem to a familiar context. Many books suggest principles to be followed in order to succeed in research or invention; some are contained in biographies of famous men, as well as in books explicitly aimed at the technical man. Some better ones: *An Introduction to the Study of Experimental Medicine,* by C. Bernard; Collier or Dover, $1.50. *Louis Pasteur: Free Lance of Science,* by R. J. Dubos; Little, Brown, 1950, $6.00. *The Psychology of Invention in the Mathematical Field,* by J. Hadamard; Dover, 1954, $1.25. *The Art of Scientific Investigation,* by W. I. B. Beveridge; Vintage Press, 1962, $1.25. *From Dream to Discovery: On being a scientist,* by H. Selye; McGraw-Hill, 1964,

$7.00. The case of the floppy-eared rabbits (described in the Hyman-Anderson essay) is discussed in a collection of papers: *The Sociology of Science*, edited by B. Barber and W. Hirsch; Free Press, 1962, $9.00.

Part II

7. The Growth of Ideas

The two studies mentioned in this chapter, performed under the auspices of the National Academy of Sciences, are available at the Clearinghouse for Scientific and Technical Information: Write to CFSTI, Department of Commerce, Springfield, Virginia 22151. The study by Harvey Brooks's committee is entitled "Applied Science & Technological Progress; a report to the Committee on Science and Astronautics, U.S. House of Representatives, by the National Academy of Sciences"; 1967, $3.00. The study by Morris Tanenbaum's committee, the Materials Advisory Board, is entitled "Principles of Research-Engineering Interaction"; 1967, $3.00. For more on Elting Morison's inventors of the nineteenth century: *Men, Machines, and Modern Times*, by Morison; MIT Press, 1966, $6.00; the author will tell you what happened again and again when technical developments were introduced: the hostility that those developments encountered. For more on the innovation process, see the essays and references in Part III of this book, particularly "From Research to Technology," by Morton. Also: *Technological Innovation: Its Environment & Management*, a report by the U.S. Department of Commerce Panel on Invention and Innovation; write to Government Printing Office, Washington, D.C. 20402; $1.25.

8. The Fear of Innovation

In *Men, Machines, and Modern Times*, by E. Morison; MIT Press, 1966, $6.00; the chapter entitled "A Case Study of Innovation" provides relevant information on the relationship between technological innovation and social change. See also *The Rate and Direction of Inventive Activity*; Princeton University Press, 1962, $12.50; especially R. Nelson's chapter entitled "The Link Between Science and Invention"; the papers in this book range from the history of technology and its impact on society, to specific cases of technological change within corporations, to problems of measuring inventive activity. See, also, *Adjusting to Technological Change*, by G. Somers; Harper & Row, 1963, $4.50; this is a collection of eight essays that bear on the author's theme. See, also, Donald Schon's recent book, *Technology and Change*; Delacorte Press, 1967, $7.95.

9. To Promote Invention

On the subject of patents, you should be aware of the availability of the hearings of the Subcommittee on Patents, Trademarks, and Copyrights, Committee of the Judiciary, U.S. Senate; it is difficult to recommend specific documents, because the hearings span some dozen years. The publication of the Patent, Trademark, and Copyright Research Institute of George Washington University, Washington, D.C., is an excellent source of patent information; the name of this publication is *Idea.* Also: *The Law of Chemical, Metallurgical and Pharmaceutical Patents,* edited by H. I. Forman; Central Book Company, 1967, $35.00. The references cited in Schon's essay, "The Fear of Innovation" are also relevant to the general area discussed in the Barnes essay.

10. Loyalties

11. What's the Boss For?

Before you begin to collect the literature relating to this pair of articles, we suggest you first read several of the other essays in this book, for these will provide both a context *for* and supplement *to* the author's observations. In particular, read the essays in Part III. The literature in the area of "industry and the scientist" began to build more than a decade ago, and one of the first of these books is now available in paperback: *Making of a Scientist,* by Anne Roe; Apollo, $1.75; this book will be valuable for the insights it provides regarding work satisfactions of scientists. Also *Motivation and Personality,* by A. H. Maslow; Harper, 1954, $5.00; this work stresses the individual's "needs" hierarchy; higher needs than survival and safety come into play as society becomes more civilized: the need to belong, to be loved, to enjoy esteem. More recent books are indebted to these earlier works. Also: *The Human Side of Enterprise,* by D. McGregor; McGraw-Hill, 1960, $4.95; McGregor emphasizes the interdependence of the individual and his organization. Also *New Patterns of Management,* by R. Likert, and *Human Organization,* also by Likert; see earlier references for specific details.

12. Supervision

For a more comprehensive discussion of Rensis Likert's thesis, you should read his two books: *New Patterns of Management*; McGraw-Hill, 1961, $6.95; and *Human Organization: Its Management and Value*; McGraw-Hill, 1967, $7.95. The latter book also discusses human resource accounting and plans for developing the procedures for this accounting. Both of these books are written for anyone concerned with the problems of organizing human resources and activity.

Other relevant books by members of the staff of the Institute for Social Research include: *Scientists in Organizations*, by Donald C. Pelz and Frank M. Andrews, Wiley, 1966, $10.00, mentioned also in this volume in relation to their papers; *The Social Psychology of Organizations*, by Daniel Katz and Robert L. Kahn, Wiley, 1966, $8.95. *Management by Participation*, by Alfred J. Marrow, David G. Bowers, and Stanley E. Seashore, Harper & Row, 1967, $7.95; *Organization Control*, by Arnold Tanenbaum, McGraw-Hill, 1968. Another book of interest is *Scientists in Industry: Conflict and Accommodation*, by William Kornhauser, University of California Press, 1962, $6.00. Also see *The Scientist in American Industry*, by S. Marcson; Harper, 1960, $3.50. This author deals with two concepts of the industrial research organization: One is based on the traditional (or "executive-authority") structure, while the other stems from the peer culture of scientists in universities, or "colleague-authority." You should also note the previous references to the work of Pelz and Andrews, who are colleagues of Likert at Michigan. And for a well-balanced view of the subject, you should read the essays in this volume by Morton and Hoskins, in Part III.

Part III

13. Science and the Organization

The Harvard University Program on Technology and Society is publishing several essays annually: "An Experiment in Understanding: The Harvard Program Two Years After," by E. Mesthene; "On Understanding Change: The Harvard Program on Technology & Society," by E. Mesthene; "Can Science Be Planned?" by H. Brooks; "Changes in the Structure of the American Economy, 1947 to 1958 and 1962," by Anne Carter. These and other essays in the series are available at Harvard University Program on Technology and Society, 61 Kirkland Street, Cambridge, Mass. 02138. On science and social needs, see: "Science, Choice, and Human Values," by A. Weinberg, *Bulletin of the Atomic Scientists, 22* (1966); "The Complexity of Scientific Choice, II: Culture, Overheads or Tertiary Industry?" by S. Toulmin, *Minerva, 4* (Winter, 1966); "Notes on the Post-Industrial Society I," by D. Bell, *The Public Interest, 6* (Winter, 1967). "Future Needs for the Support of Basic Research," by H. Brooks, in *Basic Research and National Goals*, a report to the Committee on Science and Astronautics; published in 1965 by Government Printing Office, Washington, D.C. On "research on research," see periodic issues of *Industrial Management Review*, published by Sloan School of Management, MIT, Cambridge, Mass. 02139; the Spring 1967 issue includes "Facts and Folklore in Research and Development Management," by E. Roberts. On organizations: *The New Industrial State,* by J. K. Galbraith; Houghton Mifflin, 1967, $6.95; *Human Dilemmas of Leadership,* by A. Zaleznik; Harper & Row, 1966, $5.95. *Big Business: A New Era,* by D. E. Lilienthal; originally published by Harper, 1953; available in paperback, Cardinal edition, $0.35.

14. Basic Research in Industry

For more on the author's observations, see "Applied Science and Manufacturing Technology," by D. Frey and J. Goldman; published in *Applied Science and Technological Progress,* a report to the Committee on Science and Astronautics, U.S. House of Representatives, by National Academy of Sciences; write CFSTI, Department of Commerce, Springfield, Va. 22151; 1967, $3.00. In the same volume, see: "Cases of Research and Development in a Diversified Company," by C. Guy Suits and Arthur Bueche; the authors are the former director and current director of research for General Electric. Also see, in the same volume: "Social Problems and National Socio-Technical Institutes," by A. Weinberg.

15. From Research to Technology

For more on "the systems philosophy" as it pertains to the Bell System, read three papers: M. J. Kelly's "The Bell Telephone Laboratories," published in the *Proceedings of the Royal Society,* London A, Vol. 203, p. 297, 1950; "Strategy in Industrial Research," by J. Fisk, published in *Research Management 6,* No. 5, 1963; "The Systems Approach in Science-Based Industry," by F. Kappel, published in *Bell Telephone Magazine 42,* No. 3, 1963. The *Harvard Business Review* has published two articles that discuss the systems philosophy in more general ways, relating it to business management: "Age of Synthesis," by Culliton, October 1962; "The Manager's Job: A Systems Approach," by Tilles, January 1963. The tools and steps of systems engineering are described in detail in two books: *System Engineering,* by H. H. Goode and R. E. Machol; McGraw-Hill, 1957, $12.00; *A Methodology for Systems Engineering,* by A. D. Hall; Van Nostrand, 1962, $12.00. J. W. Gardner's *Self-Renewal — the Individual and the Innovative Society* is an eloquent statement of the way in which the systems approach can be applied to the whole field of social and human relations. Available in hard cover, Harper & Row, 1963, $3.50, and paperback, Harper Colophon Books, 1965, $1.45.

16. From Materials to Systems

Two essays set the stage for this one: One is Morton's "From Research to Technology," which appears in this volume; the other is "Microelectronics," by Hittinger and Sparks, in the November 1965 issue of *Scientific American*; the authors of this latter article are executive directors in electronic components at Bell Telephone Laboratories. An earlier article on microcircuits appeared as "Integrated Circuits," by Sprague, in the April 1964 issue of *International Science and Technology*. The most complete collection of papers in the field is E. Keonjian's *Microelectronics*; McGraw-Hill, 1963, $12.50; the book is a good cross-sectional view of the industry, written by many of the leading technical people in the field. Bell Labs' approach to integrated electronics is covered in the November-December 1966 issue of *Bell Labs Record*.

17. Science on Park Avenue

The American Management Association has published a number of reports relating to the author's subject; write to AMA at 135 West 50th Street, New York 10019; in particular, see: "The Management of Scientific Talent," edited by J. Blood, 1963, *Management Report* 76; "The Market Orientation of R&D," by M. Hanan, 1965, *Management Bulletin* 72; "The Case for Research Accountability," by McLoughlin, 1967, *Management Bulletin* 98; "Organizing the R&D Function," by A. Stanley, 1965, *Research Study* 72. See, also, the following articles from the *Harvard Business Review*: "Twelve Fables of Research Management," by P. Drucker, Vol. 41, No. 1, 1963; "How to Evaluate Research Output," by J. Quinn, Vol. 38, No. 2, 1960. Also *Technological Planning on the Corporate Level*, by J. Bright, Harvard University Press, 1962. And from two other management publications: "Challenging the Scientist," by R. Hershey, in *Dun's Review*, Vol. 87, No. 6, 1966; "Instilling Efficiency in R&D," by Sharp, in *Administrative Management*, Vol. XXV, No. 4, 1964.

18. Designing a Technical Company

The key article in the engineering approach to the firm is "Common Foundations Underlying Engineering and Management," by J. Forrester; *IEEE Spectrum*, September 1964. From there you can turn to *Industrial Dynamics*, by Forrester; MIT Press, 1961, $20.00, $12.50 in paperback; this book provides the feedback-system-simulation approach. If you really get interested in computer simulation, you will need the *Dynamo User's Manual*, second edition, by A. Pugh; MIT Press, 1963, $4.50. For an industrial dynamics analysis that has direct relevance to the problems posed in Hoskins' essay, read *The Dynamics of Research and Development,* by E. Roberts; Harper & Row, 1964, $10.95. Another interesting industrial dynamics contribution is *Resource Acquisition in Corporate Growth,* by D. Packer; MIT Press, 1964, $6.00. The psychology of management, and particularly of innovation, is more diffuse, since half a dozen kinds of professionals look at the subject from special viewpoints. The best beginning is *Integrating the Individual and the Organization,* by C. Argyris; Wiley, 1964, $5.95. The beginning section is a review that ties the literature together in a very organized way; Argyris attempts the beginning of a general theory of the subject; the question of how successful he is must be left to professionals, but the beginnings of the theory give a better structure to the field than any amount of bibliographic listings could do. Of the many other studies of the "new management" currently available, one of the most readable is *The Human Side of Enterprise,* by D. McGregor; McGraw-Hill, 1960, $5.50. (If you expect to penetrate much beyond Argyris and McGregor in psychology, you will need a good lexicon, for the jargon is about as prickly as it could be and still be in English.) For the short course in the field, read (in the order given here): Forrester's IEEE article, Roberts, Packer, Argyris, and McGregor. You should also read *Eupsychian Management,* by A. Maslow; Dorsey Press, 1965, $3.50. Maslow is a theoretical psychologist who, while generally approving the arguments advanced by our authors, is careful to point out that these tools are very much of our time and place, and that they do not necessarily apply elsewhere — in, say, a developing country.

19. How the U.S. Buys Research

A broad treatment of the management of research and development projects is provided in *The Dynamics of Research and Development,* by E. Roberts; Harper & Row, 1964, $10.95. Edward Roberts has also written a pair of recent articles that relate to the subject: "Questioning the Cost/Effectiveness of the R&D Procurement Process," in *Research Program Effectiveness,* edited by M. C. Yovits; Gordon and Breach, 1966, $29.50. Also "Facts and Folklore in Research and Development Management," *Industrial Management Review,* Vol. 8, No. 2, Spring 1967; published by Sloan School of Management, MIT, Cambridge, Mass. 02139. See, also, *The Weapons Acquisition Process: An Economic Analysis,* by M. J. Peck and F. M. Scherer; Harvard Business School, 1962, $10.00; this book discusses most aspects of large military systems management, but does not say much about R&D. In both this book and the book by Roberts, the nonspecialist should skip the mathematical formulations; aside from these, both books are easy reading. See, also, *The Weapons Acquisition Process: Economic Incentives*, by F. M. Scherer; Harvard Business School, 1964, $7.50; this is an excellent discussion of real and contract incentives in government contracting. Also "Military Procurement and Contracting: An Economic Analysis," by Moore; *RAND Memorandum* RM 2948-PR-1962; this is an advanced version of both "weapons acquisition" books, using different data and analyses. Also *The Economics of Defense in the Nuclear Age*, by C. J. Hitch and R. McKean; Harvard University Press, 1960, $9.50; the book's strength lies in the broad topic coverage of defense policy and rational analysis methods used by RAND in its early work for the Defense Department. Also *Adaptability for Survival in the Defense Industry*, edited by C. Vaughn; Boston University Press, 1965.

Index